Light
Refractions

Light
Refractions

Thomas H. Middleton

Introduction by Allen Walker Read
Foreword by Laurence Urdang

VERBATIM

First printing, December 1975.
Second printing, August 1976.

Library of Congress Catalog Card Number: 75-38232

Printed in the United States of America.

PUBLISHER'S FOREWORD

Ever since I can remember, I have been intrigued by language. Not just English, but *language*—how it functions, how it conveys ideas, meaning, beauty, love, hate, propaganda, how it can, in a few words that require only seconds to utter, hear, write, or read, describe the events of millennia, the culture of all of mankind. Language captivates me; it entrances me. Words beguile me; grammar and syntax fascinate me. Yet only recently did I discover that language has the same effect on others.

At the same time, I discovered that people have strong opinions about language: a discussion about obscenities, about *ain't,* or about "good usage" can spark emotions every bit as virulent as those evoked by controversies about truth, justice, politics, motherhood, and women's lib. Few people stand mute on the subject of language. Often (if not usually) controversial, language is an issue for all time.

One might therefore expect an enormous library to be available on language, but that is not the case. Aside from the professional books and journals on the subject—chiefly

specialized works comprehensible only to the specialist—
the public was, until recently, served by only a handful
of books, indifferent in quality for the most part, and by
the occasional features in the Sunday supplements.

But in 1972 there began to appear in *Saturday Review/
World* a regularly fortnightly column on language, called
"Light Refractions." Thomas H. Middleton, known to
SR/W and Sunday *New York Times* readers for his Double-
Crostics, exhibiting the cool detachment of the profes-
sional observer, quickly won an enthusiastic following for
"Light Refractions," and I have been one of the enthu-
siasts from the start. Like many of his fans, I clipped each
article and saved it.

When VERBATIM, The Language Quarterly, began
publication in 1974, the subscriber list was so small that
I knew every name on it; when a subscription order arrived from Thomas H. Middleton, I felt flattered: we were
gaining the kind of recognition we were seeking.

I worry a great deal about stuffiness of style in VER-
BATIM, and browsing through "Light Refractions" helps
to remind me that the subject of language, which can so
easily become bogged down in tiresome technicalities
turgidly treated, can be written about in an interesting
and entertaining way.

It dawned on me that "Light Refractions" deserved a
fate better than relegation to the pile of magazines one
saves but—with all the best intentions—never comes to
read again. Tom Middleton was kind enough to allow us
to select two score or so of his articles for collection into
this anthology, the first of what we hope will become a
series of popular books about language to be published
by VERBATIM. No association could have been more

congenial: Tom's with *SR/W*, ours with Tom, or ours with *SR/W*, to whom we wish to express our gratitude for their ready agreement to release the rights to "Light Refractions."

The collection that follows is not complete: some of the articles would have been too difficult to understand out of their topical context. But many of them are here, both for the reader's delectation and to serve as a record of some of the most readable, interesting, and important writing about language to have emerged since H. L. Mencken.

—Laurence Urdang

INTRODUCTION

Whenever we talk or write, we are making long strings of choices. Fortunately these choices are mostly made unconsciously and only occasionally do we face a problem. Mr. Middleton has a talent for seizing on such problems, holding them up for inspection, and offering enlightening comment.

In academic departments of English, these questions are dealt with in an amorphous field called "usage"—partly grammar, partly rhetoric, partly "word study," and most of all sociology. In its most prestigious form (see Mr. Middleton on the status of *prestigious*), the field is known as "socio-linguistics." This name gives an investigator the right to indulge in the trappings of scientific research, with graphs, tables, and controlled experiments.

Mr. Middleton, however, chooses to base his discussions on his rich experience as a user and experimenter with language. He is not content to reflect traditional views, as they are handed down in schools, families, and cocktail conversations. There is a large body of linguistic opinion which Leonard Bloomfield called "tribal belief." The reader will find that in the essays of this book, such beliefs are subject to independent scrutiny and re-assessment.

All of us bear some scars as a result of linguistic criticism from childhood on. For some blunder or ineptness it is likely that we have been made fun of, treated with contempt, or held up to ridicule. I dredge up the memory from the age of ten that for the word *motorcycle* I used the pronunciation "motor-sickle" (once very common, I believe), but the newer pronunciation "motor-sigh-kle" was coming in, and the whole classroom burst out laughing at me. A few years later, in a high school composition, I happened to use the word *unkempt*, and the teacher insisted that I must change it to *unkept*, inasmuch as *unkempt* was a colloquial distortion. I felt ill-treated, because I was sure that I had heard *unkempt* from good sources. I now know that the teacher was dead wrong. I learned many years later that *unkempt* is an old form related to *uncombed*, and the change of vowel from "o" to "e" represents an Anglo-Saxon sound change (the "i-umlaut") still embedded in present-day speech.

Even in adult life, the criticism of speech is used as a bludgeon to establish one's superiority. I have known of wives who hold their husbands in subjection by criticizing their speech. At a party one such wife said to everyone within hearing distance, "I hope you'll forgive my husband for his bad English," and we all looked at him pityingly. Frankly I had not noticed anything out of the ordinary in his speech, nor did I later. It is difficult to make any criticism without seeming to be superior. The troublesome word *spontaneity* can have three pronunciations of its stressed syllable—*knee, nay* (heard more often than not, a carryover from the adjective *spontaneous*), or *nigh.* A colleague of mine in an English department was unusual in using the third of these, and as gently as I could I called to his attention that the form of the word called for a *knee* pronunciation. He crumbled before my eyes, confessing his cultural insecurities, and I felt as if I had been using a bludgeon. I had meant well, but I would have much preferred to hear *spontan-eye-ity* rather than to go through such a painful scene.

Mr. Middleton is not afraid to wrestle with recalcitrant problems—those that spring from the deficiencies of the language itself. One of these is the "everyone . . . they" problem. In spite of the tremendous resources of the English language, it lets us down in this instance. Think of the sentence, "Everyone applauded, and I'm glad he did." How absurd can one get? It is comforting to find Mr. Middleton's conclusion: "My choice is to accept 'I don't think anybody knows where they are' as perfectly all right. It seems to me that to do otherwise threatens to leave us tongue-tied." Even since he wrote that, we have increased our feeling of guilt over the acceptance of "he" in the reference "he-or-she." The pedantic person who is proud of saying "everybody has *his* own pet peeves" is driven by women's lib to the clumsier "*his or her* own pet peeves."

The human dimensions of the *ain't* problem are recognized in these essays. I can dredge up another memory from the age of ten. My sixth-grade teacher asked the class members to watch each other's speech and to make a list of all the mistakes they might hear. Some anonymous member, perhaps out of spite, listed that I had been heard to say the word *ain't.* Already I was so indoctrinated against the word that I would have bitten off my tongue

before using it. I went to the teacher with trembling voice
and tears in my eyes to assure her positively that I could
not have said it. Though she could not change the record,
she did her best to comfort me. I also remember metaphysical arguments on the playground about whether *ain't*
was in the dictionary. I say metaphysical because we
never thought of looking the word up, though dictionaries
were readily available; but we preferred to treat it as a
speculative matter. Such a background helps to explain
how *ain't* is a signal word. If one uses it, one is signaling
one's choice of a vulgar, illiterate posture, while avoidance
leaves one among the right-minded defenders of the tribal
culture. You will note Mr. Middleton's paradoxical explanation later on.

The wide range of these essays is illustrated by the con-
trast between "Smashing" and "Language and Survival."
The ubiquity of *smashing* in British colloquial speech has
been a marvel to behold. The agent form *smasher* is docu-
mented in the highest quarters. In a letter of February
15, 1947, Princess Elizabeth, according to her nanny's
book, *The Little Princesses*, wrote as follows: "The officers
are charming and we have had great fun with them. . . .
There are one or two real smashers and I bet you'd have a
wonderful time if you were here." The Princess has now
become the titular guardian of "The Queen's English,"
but even the royal sanction has not brought this word into
American currency. We can agree with Mr. Middleton
that such a word has little chance. This is a trivial matter,
but the larger context of language may well determine
our survival. It is heartening to learn of Mr. Middleton's
optimistic appraisal: As he says, "the average American
speaks better English, with a much larger vocabulary

and far fewer serious grammatical errors, than ever before."

In every discussion of usage, not far below the surface is the question, "Whose usage?" In cultures that have a well-recognized élite (sometimes formalized into an "academy" for linguistic matters), the usage that has prestige is readily discernible. The problem is a difficult one in countries with great social diversity. The academic world will not suffice, as it is too often outside the mainstream. "Old money" versus "new money" has sometimes made a difference. The "guardians of culture" are usually self-appointed—those who make themselves heard by crying out against "verbal atrocities," "corruptions," "ugly" words, "encroaching barbarism," or "deterioration." Mr. Middleton does not belong to this simplistic school. He combines a high-spirited interest in language with an open, sunny approach that recognizes the complexities of the subject but that brings a sober wisdom to the solution of its problems. These essays are an adventure in wholesome linguistic understanding.

—Allen Walker Read
Columbia University

Light
Refractions

WORDS, WORDS, WORDS

> POLONIUS: What do you read, my lord?
> HAMLET: Words, words, words.
> POLONIUS: What is the matter, my lord?
> HAMLET: Between who?
> POLONIUS: I mean, the matter that you
> read, my lord.

Had Polonius been a logophile, he'd have been satisfied with Hamlet's "Words, words, words"; he wouldn't have inquired what the matter was; and Hamlet wouldn't have said, "Between who?"

A logophile is a lover of words, of course. You won't find logophile in any dictionary I know of, but I needed a word meaning "a lover of words" and that's the best I could do.

There's "logomaniac," defined in the *Oxford English Dictionary* as "one who is insanely interested in words." But not all lovers of words are insane about them; ergo, logophile, or, to compound logodaedaly ad absurdum, ergologophile.

A logophile is one who enjoys reading dictionaries, who loves words for their own sake, for their sound and their sense (or even sometimes their nonsense), for their background and upbringing. For a logophile, words all by themselves can move and excite. Words untouched by a poet—words simply as they are in their raw, unpolished, dictionary setting.

Polonius was wordy, God knows. Had he been not merely wordy but a lover of words, the conversation might have gone something like this:

POLONIUS: What do you read, my lord?
HAMLET: Words, words, words.
POLONIUS: "What, good my lord! A fellow logophile!
A fancier of lexicons! to know
The joy of tasting nouns and verbs, to love
The word itself, unmindful of the matter
Signified, or, mindful, caring mostly
For the sound, the felt relationship
Between th' expression and the thing expressed.
We epicures of words, my lord, we gluttons
Of the glossary are privy to a
Love so esoteric, so removed
From base corporeal lust for tangibles
And on and on and on and on and on

By this time the crowd in the pit would have started throwing things.

Shakespeare knew what he was doing.

Words can be sensuous. That's one of them: sensuous. A perfect word, it moves sensuously in the mouth. It's a champagne of a word. Sensuous is sensuous, no tautology intended. There's a certain onomatopoeia, as there is in "chaotic." Chaotic *sounds* chaotic.

Some words are contra-onomatopoeic. The first time I was conscious of a word that didn't sound right for the thing it represented was when I first studied plane geometry and got my compass and protractor. I felt strongly, and still do, that the compass should have been called a protractor and the protractor on the other hand should have been called a compass.

Here's a gadget with one long steel leg ending in a sharp point and one short steel leg ending in a socket into which you clamp a stubby pencil, pegleg-like. The legs can be spread apart or moved close together or locked in position. Fix the point in a sheet of paper and swing the pegleg to make a circle. Make a design of bigger and smaller circles. Cover the paper with them. A machine like that deserves to be called a protractor, and that little thing that just lies there looking semicircular doesn't. I didn't speak up about it in those early days. I'm sure it's too late to do anything about it now.

Probably the most notoriously contra-onomatopoeic word in the language is "fulsome," which is why it's so often misused. In its early history fulsome used to mean "copious, abundant, plump." Somehow it came to mean "gross, disgusting, nauseating, tasteless." But fulsome sounds rich, flowing, billowing, wholesome(?). Some day, I hope, it will mean all that. It will have been misused often enough to become the word it should be.

Misuse, of course, is one of the great forces for changing a language. If the dictionaries say one thing and enough people say something else, guess who's going to have to give way. There will always be a small army of linguistic conservatives, respected, correct, and valiant, standing on the sidelines calling "Foul!" I usually find myself among them. But ultimately we're doomed, and I suppose it's just as well that it is that way. Language is alive, and change is of its essence.

July 4, 1972

A SLIGHT
CHINESE ACCENT

Now that President Nixon has given official sanction to the heretofore subversive notion that China is China and Taiwan is either Taiwan or Formosa or both, and now that signs of Sinomania, or the China-craze, are everywhere about us, I have looked again at my favorite China reference: *The Usage of English Slang and Colloquialism.*

I first heard of the booklet on a movie set in Taipei. We were shooting Robert Wise's film *The Sand Pebbles.* I had been hired as dialogue coach, and we'd engaged some of the local Chinese and Taiwanese college students to help in the various departments—wardrobe, make-up, etc.

A lovely young wardrobe assistant, a Chinese girl of great delicacy and decorum, suddenly asked me, "Hey, Tom, where you betta half?" I blinked and countered with "What?"

"Where you excess baggage?" she replied.

"Kitty," I said (Kitty was her English name; I can't remember her Chinese one, which was more suitable), "where did you learn those expressions?"

She beamed and showed me her copy of *The Usage of English Slang and Colloquialism,* a paperback book of 112 pages. I thumbed through it, asked her where she got it, and that evening I bought half a dozen copies, all but one of which I've given away. I've seen it reduce jaded comedy writers to tear-be-slobbered laughter in five minutes even or less.

I didn't tell Kitty, and I'm glad I didn't, but there's scarcely an expression in the entire thing that isn't somehow flawed, in some cases outrageously.

The book lists expressions in English, followed by their translations in Chinese ideograms, with examples of their use in context. "War paint," for instance, is used in the sentence, "She is too old to be charm, so she put too much powder as in war paint."

One of my favorites is "To be a ladies' man": "The vicar is likely to be a success; you see he's such a ladies' man."

I think I got to be pretty good at spotting the Chinese who had studied *The Usage, etc.* They had a special flair with language. Harry, the super-salesman at George's Tailors, was a case in point. George's was an establishment famous for its ridiculously low prices. The clothes seldom fit properly, but the prices were too attractive to pass up. The bad fit didn't matter a great deal, because George's garments were sewn together with a thread that dissolved during or immediately after the first washing or trip to the dry cleaner.

We never met George. The head man as far as we knew was Harry, a tall young Chinese whose language was salty and whose style was colorful. He was a *Usage of, etc.* trainee.

My wife was fitted for a *chi pao* (Mandarin for the high-necked, close-fitting Chinese dress called a *chong sam* in Cantonese), and Harry assured us it would be ready in time for a party Friday night. "No sweat, no sweat," he insisted.

When Jeannie and I went in to pick it up on Friday, Harry slapped his palm to his forehead, crying "God bless my soul! Hold your horse, I go see the tailor!" He dashed out the side door and down the alley.

It was our thought that the tailor was, in fact, anyone who could pedal a sewing machine at the time. We pictured the machine as a treadle-operated antique hung with their own spun-sugar thread.

Harry was gone for some fifteen minutes while we wandered about the shop fingering bolts of material, much of which was excellent. Finally he returned, greatly agitated. "I so pist off," he said. (That's how it's spelled in the book. A four-letter word.) "Not ready yet. Nother hour, you come back. Goddam tailor!"

The *chi pao* was ready an hour later, and it was beautiful on Jeannie, though it cut off all circulation at the shoulders so that her arms quickly became numb. That didn't stop her from wearing it, of course. The seams lasted through the second dry cleaning. Not bad.

In what we can hope will be one of the great achievements of the 1970s—the closing of the gap, the opening of the doors, the meeting of East and West—*The Usage of English Slang and Colloquialism* may well contribute its share of good cheer. If a Chinese acquaintance said to you, "God bless my soul! Who would have thought your behave is like that," or, "He is nothing but a chip of old block," it could make your day, and now you'd know where the slang and colloquialism came from.

Perhaps some Old China Hand, some former member of the U.S. Navy who knew the Yangtze in the days of gunboat diplomacy and picked up a sailor's knowledge of Mandarin, will write "The Usage of Chinese Slang and Colloquialism" for Americans. People-to-people contact between the two nations could move with the zest of a Marx Brothers romp, and who could fault that kind of Marxism?

July 18, 1972

ᒣᒣᒣᒣᒣᒣᒣᒣᒣᒣᒣᒣᒣᒣᒣᒣᒣᒣᒣᒣᒣᒣᒣᒣᒣᒣᒣᒣᒣᒣᒣᒣᒣᒣᒣᒣᒣᒣ

WHAT IS THE OPPOSITE OF "POSTPONE"?

I was asked that question by Professor Louie J. Fant, Jr., author of *Say It With Hands* and *Ameslan, an Introduction to American Sign Language.* He showed me the sign for "postpone." You make circles of your thumbs and forefingers and move them away from your body, as though you were moving pegs a few spaces on a large cribbage board. "Now," he said, "suppose you wanted to change the date of a football game from October 15th to October 14th. You sign it like this." He moved the imaginary cribbage pegs toward his body. "That's the opposite of postpone. How do you say it in English? Is there a word?"

I don't think there is. There's certainly no such word as "prepone." The best way to say it is "move ahead": "The game has been moved ahead from the 15th to the 14th."

That's right, but somehow it sounds all wrong, especially if you've just seen the same meaning signed with a clearly logical backward movement of the hands indicating a greater closeness spatially and, by inference, temporally. Moving the hands forward indicates, clearly, "postpone."

That was the beginning of my friendship with Lou Fant and of my sudden, belated interest in the deaf and the subject of non-verbal communication.

Lou Fant was born to deaf parents. His father had been deaf since birth and his mother had lost her hearing when she was about two years old. Lou learned to speak and to sign simultaneously as a baby. His aunts, uncles, and grandparents taught him speech as his parents taught him to sign.

"Signing" and "using signs" are the terms most often used for that largely non-verbal process. "Speaking sign language" is considered inappropriate, as it implies speech and therefore words.

I'd never really imagined total deafness. Seeing groups of young people using sign language with one another, I assumed that what I saw was comparable to semaphore or some other code—that while one was signaling with his hands, he might be thinking, "I'll meet you tomorrow at the Automat at 12:30, O.K.?" and that his friend was reading the signals, the words registering in his mind, "I'll meet you tomorrow at the Automat at 12:30, O.K.?" and that he might respond with his hands and his thoughts, "O.K. Bring your sister."

But if you've been born deaf, you can't possibly think in language as hearing people think in language. You think in images, symbols, and, if you're an unusually well-educated deaf person, in printed words, signed words, or lip-read words, but always in *visual* language. For the deaf, there's no translation from sign to word to meaning. Words have nothing to do with it, for the most part.

It's self-evident that a deaf person can't appreciate

music; it's less apparent that, with rare exceptions, the deaf can't appreciate poetry. Poetry must be heard, at least in the mind, and one who has never heard can scarcely imagine the sound of words. There is no meter or rhyme on the printed page. It requires a hearing mind to midwife the poetry. It's true that there are a few extraordinary people who, though born deaf, have managed to embrace and even to create poetry, but they are not only atypical, they are among the deaf what Shakespeare, Milton, Bach, and Beethoven are among the hearing. (I include Beethoven here advisedly. Though he became deaf, he must, in this context, be numbered among the hearing, because he *knew* the sound of music.) For the average deaf person, dance is as poetry; light and color are as music.

For three years, Fant was connected with the National Theater of the Deaf, which toured the country presenting plays for the deaf, with vocal narrations for the hearing. Through this association, he got interested in the theater in general and was persuaded by a theatrical producer to go into acting. He's now working in films and television when he's not teaching or writing.

He teaches Ameslan (a word he coined for American sign language) at California State University at Northridge to eager students, both deaf and hearing. It's an illuminating experience for the deaf and the hearing to work together in plays they adapt for both the deaf and the hearing—using words and signs, dance and mime.

Though the United States is ahead of the rest of the world in its care for and understanding of the deaf, we have a long way to go. Deafness is an invisible handicap and one of the least understood. Professor Fant's students give him great hope. Those who hear are learning the language of the deaf, establishing rapport and even empathy. They're working in a dramatic relationship, both on-stage and off.

And if you think those who hear have nothing to learn from the deaf, give some thought to "the opposite of postpone."

September 26, 1972

BANDERSNATCH IN THE OUTFIELD

T he manager three-inninged his pitcher."

That was on a "Now let's turn to the World of Sports" segment of a news broadcast. The sportscaster mentioned the names of the manager and the pitcher, but the verb was so startling as to eclipse them from memory. Nor can I remember the name of the sportscaster who coined the term **three-inning**, *v.t.;* **1.** To let (someone) play for three innings. **2.** To remove (someone) from a game after three innings.

This opens the curtains on unexplored areas of logodaedaly. We could two-hour someone. We could two-minute, or short shrift, undesirables. We might even Tuesday and Thursday one another.

Making verbs out of nouns is nothing new to the sports-announcing dodge, as anyone who sits glassy-eyed before the idiot-box during the football season knows. We all know how tough it is to defense against Fran Tarkenton and Roger Staubach. The reason is that they not only can throw the bomb but they're great scramblers.

Language tends to get more colorful in sports reporting

than in most fields, with good reason. How often can you say, "Frumious hits a high fly to center (or right or left) field, and Bandersnatch catches (or misses) it!" You'd lose the ear of the most ardent fan. The language has to lend new color to acts that are in themselves colorful but, because of the structure of games, repetitious. So Frumious belts, clouts, lifts, blasts, swats, whops, wallops, slaps, or sends the ball high into center, left, or right; and Bandersnatch gets it, traps its, spears it, snags it, shags it, gathers it, or hauls it in.

In football a ballcarrier who doesn't make a touchdown is almost always tackled or run out of bounds. Usually tackled. On the air he is smeared, clobbered, ganged, hit hard, brought down, buried, stopped, stomped, or crushed.

Back in the late Forties I was listening to a football game on the radio. It might have been Army–Notre Dame, I don't remember. All I remember is some of the announcer's phrases. No ball-carrier should have survived that day, but they did. Runners were shot down, riddled, ambushed, rifled, gunned, shelled, and bombarded. To hear it on radio, the field was covered with ordnance and corpses.

This same announcer told us that at the start of the season one of the coaches—say it was Red Blaik—had been down-hearted. What he actually said was, "Red's spirits were lower than a midget's heels in the Holland Tunnel," the reason being that his team consisted almost entirely of rookies. As he put it, "They were greener than a sick Irishman in the grass on St. Patrick's Day."

How's that for color?

During the Sixties there was a peculiar sports cliché that dealt with "good hands." If a player made a spectacular catch of a ball (foot- or base-), it was because he had good hands. Luis Aparicio hears the crack of hickory on horsehide (a cliché from the Golden Age of Sport). He moves like a startled rabbit to his right, twists his body with instant instinct, catches the ball in the glove on his left hand, shifts to his right hand, shifts his weight to his right foot, cocks his arm, and throws with brilliant speed and accuracy to the first baseman for a put-out. It's one uninterrupted motion—an instance of dazzling athletic co-ordination. You see it on the tube, and in your own way you register something like "Zowie!"

The guys at the microphones say, "A great play. Aparicio has good hands."

"Great hands."

"Right. Aparicio has great hands."

Rubinstein has better hands, but I wouldn't put him at shortstop.

October 24, 1972

L'INVASION AMÉRICAINE

Paris

The interchange of words between French and English goes back at least to 1066, and there have been some curious volleys back and forth during the intervening centuries.

Take a simple one: Beef, of course, comes from the Old French *boef.* In English it's been called *boef, bouf, befe, byffe, boeff,* and a few other Chaucerian-looking things, as well as beef. In modern French it's *boeuf,* but the French have borrowed back the English beef in the peculiar culinary form of *bifteck.* Years ago I remember seeing on French menus not only *"bifteck de boeuf,"* but *"bifteck de cheval."*

And the game continues.

For decades the old-guard French *académiciens* have been decrying the infestation of their language by Anglicisms and the even more diabolical Americanisms, which appear to threaten not only the tongue but the whole body of French society. And now the dread *Coca-Culture* seems to have established an invulnerable beachhead, and the *vieux gars* must be gnashing their *dents*, wringing their *mains*, and tearing their *cheveux* out by the roots every time they see a Le Drugstore—that *affreux* chain of emporiums where any *français* who wants to lower his standard of eating can get a facsimile of standard American lunch-counter fare. Le Drugstore has places all over Paris, and *Parisiens* flock into them. They have most of the features of the traditional American drugstore: toasters, records, books, Seltzer bottles, radios, even a small drug department. But the overwhelming emphasis is on tasteless food. I doubt that they're a serious threat to *haute cuisine* or even to good plain French cooking, but they certainly make it easier to find a bad meal in Paris, if that's what you're looking for.

While we're on food, a rather wacky instance of cultural exchange has struck me here: Paris is full of pizzerias these days, and while it's only conjecture on my part, I think that it *must* be an American influence. The pizza has crossed the Franco-Italian border via the U.S.A., or I'm *l'oncle d'un singe*.

An outstanding linguistic incursion by the English or the American into French is the word "self," which has recently almost explosively burgeoned.

Everyone knows that there are now self-service *supermarchés* in every major French city, but I've been surprised by the ubiquitous novel uses of "self-." "Self-service" and "do-it-yourself" have been truncated and amalgamated into "self-," so that here you find *Self-Marché* and *Self-Snack*, and yesterday I saw a small screw driver whose card-board-and-plastic package billed it as a *"self-outil."* A self-tool!

Perhaps the most alarming use I've seen of "self-" is on a butcher shop that calls itself a *Self-Boucherie*. To my Anglo-Saxon mind, that brought up a series of macabre visions starting with Vincent van Gogh.

In at least one instance, "self" stands by itself. Many of the French towns situated along the main highways have billboards erected by their *syndicats d'initiative* heralding the features offered by the towns if motorists will only stick around awhile. I didn't see this, myself, but a *cher ami*, George Gonneau, who is a resident of Paris and whose *véracité* is *impeccable*, says that one of these towns has a sign advertising SA RIVIERE SES PARKINGS SON SELF!

When *thés dansants* were popular in England, *tea dancings* were popular in France. Obviously, a *thé dansant* had more *chic* than a tea dance north of the Channel, and a *tea dancing* had a tonier *ton* on the Continent. The use of an exotic term always lends a little class, so there's a store in Paris that sells *parfums* and *eaux de toilette*, and the management evidently decided *toilette* didn't have quite the *elan*. They call their shop TOILET CLUB. No, I didn't make that up.

One of my pet anomalies is a *boîte* on the Champs Elysées called *Frenchie's Saloon*. It's a *petit coin du Wild-Ouest* in the *huitième arrondissement*.

Ya see, thishere French fella blew inta Paris an opened thishere dance hall. . . .

And still it's Paris. *Plus ça change.* . . .

November 21, 1972

A PROPER HORSE RACE

A recent *Light Refractions* on sportscasters' talk rang a few bells, apparently, and some of the jingles got back to me.

Richard Jeffries, who says he's an avid fan of all spectator sports, writes, "When the first lion charged across the Colosseum and sank his claws six inches into the first Christian martyr, then munched off fourteen inches of arm and shredded twenty inches of leg, a fat Roman senator pontificated, 'It's a game of inches.' If one more football announcer tells us that it's a game of inches, I'm going to give up watching, much as I need the exercise."

A sentiment widely shared.

I'd begun to suspect that anyone who announced sports had to pretend he knew nothing of grammar: "He can run the ball good"; "He hits pretty good"; "He's th'own' rill good this year." But Mr. H. A. Wever tells me there's a sportscaster in St. Louis named Jack Buck who uses words like "exquisite"; whose grammar is beyond cavil; who "has overcome the handicaps of literacy, erudition, and wit in attaining at least provincial success in his field."

May his tribe increase.

Robert Van Scoyk points out, rightly, that what gives sportscasting its individuality is not necessarily the words. He mentions Clem McCarthy. Clem McCarthy could have eliminated words altogether and still glued you to a horse race or prize fight. He sounded like a rusty lawn mower destroying a gravel path. No matter what the action, he gave it a rip-roaring urgency.

I had thought of Clem McCarthy on my own, in fact, when I heard a British broadcast of a horse race a few weeks ago. Like thinking of a hamburger with onion while eating a watercress sandwich.

If I had a tape of the English horse race, I'd guarantee the accuracy of the following, but as I haven't, I'll do what I can to approximate.

First, the voice. Think of the last English film you've seen in which the Scotland Yard inspector, his back to the man *we* know is the killer, runs his finger along a table top, scanning it for clues, and says with nasal insouciance, "No, we've nothing definite to go on as yet. Still—we'd be most grateful if you didn't leave the country. We'd like to keep in touch." You have the offhandedness? Okay. In that tone of voice here's the horse race.

"Lovely day here. The sky a cerulean blue. [*Pause*] Just a lonely cloud or two to our left. [*Pause*] That would be the East. [*Pause*] A brisk breeze. Just to refresh, but not to chill. Ah, the race has started. The horses are still rather bunched up. Surprisingly few people out here on such a lovely day. I suppose it has to do with the morning overcast. Difficult to predict the weather would turn so glorious. Blue Boy seems to be in first place now. It looks like

Lucky Penny second and Bobbin third. Bobbin, of course, is the horse that was trained as a jumper and didn't begin flat racing until a year ago. He's fared rather well in the past six months. Exciting horse. Those one or two clouds I mentioned in the East have been joined by several others. A lovely formation. Like a collage of cotton wool. It could, of course, turn to nasty weather. No way of knowing, this time of year. Still. Ah, Silly Billy has won. Came from behind. Jonquil second. Sweetbrier third. Jolly good race."

It brought to mind the scene from *My Fair Lady*—"What a thrilling, absolutely chilling running of the Ascot opening race." And, of course, Clem McCarthy.

December 19, 1972

AUTHORITIES' TALK

D aisy Ashford, the English author, was born in 1883. She wrote her masterpiece, *The Young Visiters,* at the age of nine, and, as Mrs. James Devlin, she died in 1972 in England.

I don't think Miss Ashford ever had anything else published. She didn't have to, of course. At age ten she could rest on her laurels, a retired novelist.

I first read *The Young Visiters* about twenty years ago when a friend lent it to me. Doubleday has just reissued it, and I've just reread it and reloved it.

Call it free association, stream of consciousness, whatever you like; the mind takes some crazy bounces, and mine caromed off *The Young Visiters* to Authorities' Talk. Authorities' Talk is the language used by the police and the military on television. Not on the dramatic shows, but on the news shows. There's a big difference. In this case my stream of consciousness shot the rapids to police talk, specifically.

Miss Ashford wrote, "Welcome sir he exclaimed good naturedly as Mr Salteena alighted rather quickly from the

viacle and please to step inside." Later, in the chapter entitled "The Crystal Palace": "Mr Salteena stood bag in
hand in the ancestle hall waiting for the viacle to convay
him to the station."

It was Miss Ashford's use of the word *viacle* that did it.
I don't know how widespread the custom is, but here in
southern California, a cop would join the American Civil
Liberties Union before he'd use the word *car* on a news
broadcast. Or, for that matter, before he'd use the word
person, except in the phrase *person or persons unknown.*
It's *individual.*

TV Reporter: Sergeant Truncheon, you were near the scene
of the robbery. Could you tell us what happened?

Sgt. Truncheon: I perceived a vehicle in front of the bank.
It appeared to be illegally parked. Two individuals
exited from the bank and entered the vehicle. They appeared to have ladies' nylon stockings drawn over their
faces, and I perceived that both individuals were carry-
ing firearms. (It's never *guns*—always *weapons* or *fire-*
arms.)

TV R: I see. What did you do?

Sgt. T: My partner and I gave immediate pursuit in our
vehicle, and we pursued the suspects' vehicle at speeds
up to one hundred twenty miles per hour.

TV R: Through the streets of downtown Los Angeles?
You don't feel that that posed an unnecessary danger to
innocent bystanders?

Sgt. T: Negative, sir. Our siren gave fair warning. The
suspects' vehicle collided with six or seven individuals
on the sidewalk, but . . .

Negative for *no* is standard, though I've never heard *positive* for *yes*.

I don't know why this is the language used by our police. Do they learn it at the police academy? Do they use it among themselves? With their wives?

Wife: More Cheerios, dear?
Cop: Negative, sweetheart. I have to get to my vehicle and apprehend those individuals committing felonies.

I asked Joseph Wambaugh, the Los Angeles policeman and author of *The New Centurions* if he had a theory as to why the police talk this way. He feels that, as the police are a paramilitary organization, they follow the lead of the military. Well then, why do the military talk this way? Perhaps the reasoning behind this mechanical language is that it sounds impersonal, therefore objective, therefore evenhanded. It sounds as evenhanded as a guillotine. I don't like it. It has an unreal quality, like the Pentagon's *hardware* and *body count* and *protective reaction*. And, of course, *delivery systems*. We've spent billions to develop what sounds like a super United Parcel Service but is in fact the most colossal complex for inflicting death and destruction in history.

While we're dealing with humanity, let's use human terms and even inhuman terms, not non-human terms. Paradoxically, there's no other word I know of for a human being that robs him of his individuality—his personal uniqueness—more than *individual*. Person, man, woman, child, boy, girl, lady, gentleman, bum, slob, human being; they're all warm-blooded. Individual is not.

As I said, it's a crazy hop from Daisy Ashford to this.

Incidentally, Miss Ashford has a policeman who sounds pleasant enough, but I wouldn't ask him for directions unless I had a bit of the necessary: "Outside Mr Salteena found a tall policeman. Could you direct me to the Crystale Palace if you please said Mr Salteena nervously.

"Well said the geniul policeman my advice would be to take a cab sir."

A nice cop and a nice cop-out. And I like his respectful presumption of Mr Salteena's affluence.

I guess one of the worst things about Authorities' Talk is that there's nothing geniul about it.

January 2, 1973

SEARCHING FOR
THE MEANING

It was inevitable that I'd buy *The Compact Edition of the Oxford English Dictionary,* but seventy-five dollars is seventy-five dollars, so I procrastinated until one day, while browsing in my local bookstore, I responded to a reckless impulse and indulged myself. After all, for the seventy-five dollars, they include a free magnifying glass.

I got home, tore off the plastic covering, and, magnifying glass in hand, looked up *weasel.*

Why, you may ask, would anyone rush to look up *weasel* in the world's greatest dictionary? The reason goes back a few years to the time when my brilliant niece, Elizabeth, then age nine, returned from England where, among other things, she'd learned the true meaning of "Pop Goes the Weasel." I recalled, as a child, having taken the phrase as meaning what it said. I imagine millions of other children had the same picture as I did of a very small weasel disappearing with a "pop," as in an animated cartoon. One moment there's a weasel, then there's a pop and a cluster of radiating lines, and then there's nothing. "Pop!" goes the weasel.

Not at all, said Elizabeth. A weasel was a tailor's iron, and *pop* was slang for *pawn*. The song was about an alcoholic tailor who went:

> Up and down the City Road,/In and out the Eagle,/That's the way the money goes—/Pop goes the weasel!/A penny for a spool of thread,/A penny for a needle,/That's the way the money goes/Pop goes the weasel!

The Eagle, she said, was a pub, and, to keep himself in ale, or vice versa, the tailor had to pawn his iron. And that's why I looked up *weasel*.

TCEOTOED says nothing about a weasel's being a tailor's iron. It does say that a weasel is not only a carnivorous animal, *Putorius nivalis,* but a native of South Carolina.

Puzzled by the omission of "a tailor's iron" under *weasel,* I turned to *The Oxford Dictionary of Quotations,* hoping it would quote the song and supply a footnote.

I expected to find "Pop Goes the Weasel" under "Anonymous." It's not; it's under "W. R. Mandale, nineteenth century." There's no explanation of its meaning, so I turned to *Bartlett's,* where it's under "Anonymous" and has the following footnote:

> The weasel was a hatter's tool, and "pop" was a term meaning to pawn or "hock." The Eagle was a music hall in the City Road. The song is attributed to W. R. Mandale.

I was ready to accept the fact that a weasel was a hatter's tool, though that is disturbingly vague, but Elizabeth's father, John May, who was once an English schoolboy, tells me that every English schoolboy knows that a weasel is a tailor's iron. Presumably the editors of TCEOTOED know it, too. Why did they leave it out?

I think *Bartlett's* also is mistaken in calling The Eagle a music hall. Nobody would call a music hall The Eagle. The Eagle was a pub.

I've looked for W. R. Mandale in every reference work I could think of and have found nothing more about him, so I feel that I can tell the truth about him without fear of contradiction: W. R. "Handy" Mandale was a music-hall performer. (This was a factor in *Bartlett's* confusion). In 1884 he was appearing at The Meadowlark, a music hall. He'd been fitted by his tailor for a new pair of checked trousers he needed in his act. He picked them up just before he was to go on-stage. When he put them on in his dressing room, he found that they were six sizes too large. "Lor' love a duck!" he exclaimed. "That Bloomin' 'Arry must of bin more'n 'arf seas over w'en 'e frew these kicksies togever!"

HE HAD to wear the trousers in his act, so he wrote a new routine in his dressing room, by way of explaining them. His appearance was greeted by great laughter. Mandale waited it out, then launched his new number. "My ty-ler, 'Arry Krishner," he said, "mykes clothes for me an' my bruvver. 'E myde these kicksies f' bofe of us. But seriously, folks, I fink 'e's ruddy sotted 'arf the time. 'E's forever poppin' 'is weasel for a few bob so's 'e kin wrap 'is filthy lips 'round a bit of alky in The Heagle, a pub just down the City Road from 'is hown hestablishment . . ." and so on until he got to his grand climax, the new song about the alcoholic tailor. "Pop Goes the Weasel!" was an instant success.

Thomas H. Middleton

How fortuitous was the relationship between "Handy" Mandale and Harry Krishner, the tailor! It gave us this great all-time favorite song, and Mandale was the prototype for the classic baggy-pants burlesque comic.

January 16, 1973

SYNDROMES, PARAMETERS, AND SIBILANTS

Some time ago, I called attention to the "good-hands syndrome" in sportscasting. If an athlete makes a brilliant catch, the sportscaster is sure to tell us that he has good hands. If you're not a football fan, you're probably going to have trouble believing this, but during the past football season, millions of viewers saw a touchdown committed on their TV tubes, and as the player scampered nimbly into the end zone, the announcer said—ready? "He has good feet!" No, I'm not kidding. Seabiscuit and Citation, of course, had good hooves.

I mentioned the "good-hands syndrome." *Syndrome* has become a fashionable word only in recent decades, though it's been around for years. Originally, of course, it was a medical term, but, according to my compact edition of the *Oxford English Dictionary*, through which I love to browse, as long ago as 1651 a man named Biggs wrote of "a farraginous Syndrome of Knaves and Fools." I can't find any mention of Biggs anywhere else. That may be the only phrase he ever wrote. It's enough. Today, we're hip-deep in syndromes.

Parameter is currently a very popular word, particularly among bureaucrats. I think it's generally used as a mispronunciation of "perimeter." A typical sentence from, say, a TV interview might go, "We've bent every effort to incorporate those subfunctions within the parameters of our organizational structure." As *perimeter*, that might make sense, perimeters being boundaries. I was puzzled as to why so many bureaucrats had decided to pronounce perimeter as parameter. I wondered if perhaps parameter meant nearly the same thing. Maybe a parameter was like a perimeter only better. I looked it up. Webster's New International, third edition:

> pa•ram•e•ter . . . 1: the relative intercept made by a plane on a crystallographic axis, the ratio of the intercepts determining the position of the plane 2a: an arbitrary constant characterizing by each of its particular values some particular member of a system (as of expressions, curves, surfaces, functions) <a ~ is a quantity which may have various values each fixed within the limits of a stated case or discusion—T. F. Weldon>; *specif*: a quantity that describes a statistical population <a clear distinction should always be drawn between ~s and estimates, i.e., between quantities which characterize the universe, and estimates of these quantities calculated from observations—*Statistical Methods in Research & Production*> <estimation of the values of the ~ s which enter into the equation representing the chosen relation—Frank Yates> b: an independent variable through functions of which other functions may be expressed <four ~ s are necessary to determine an event, namely the three which determine its position and the one which determines its time—P. W. Bridgman>

I still don't know exactly what a parameter is, but I don't think it's something within which we can incorporate anything. There's no within which about a parameter.

I suspect that the first person to use the word *parameter* in public probably knew what he was talking about. He might even have been a crystallographer. But I'm sure I've never heard the word used properly, and I've heard it used dozens of times. I suppose that some day it will mean perimeter, only better.

Two words that have been brutalized with increasing frequency in recent years are *accessory* and *succinct*. For some reason, the two *c*'s are often pronounced as though they were *s*'s, or perhaps one *c*. They come out "assessory" and "sussinct." I find it disturbing. Webster's Third admits the possibility of "assessory" and "sussinct" (as well as "flassid" for *flaccid*), but notes that, at least in the case of "assessory," it's chiefly substandard. None of my older dictionaries mentions these bastard pronunciations at all; nor does the Oxford. So we can assume that the custom started during the past two or three decades. How? Who would arbitrarily decide to pronounce two *c*'s as one, and why would anyone follow suit? Accessory is rooted in accede. I've never heard accede pronounced "acede." "Assessory" sounds related to assess. Presumably, at some point in time, some sibilant-loving, anti-velar simpleton decided "assessory" had more *tone* than *accessory*. The villain must have been someone people listened to and even respected and, unfortunately, emulated. I favor public contumely for anyone guilty of "assessory" or "sussinct" or "flassid," lest some day we be plagued by "susseed" and "assident"; and of course they'll be legitimized. After all, nothing susseeds like sussess.

February 13, 1973

▮▮▮▮▮▮▮▮▮▮▮▮▮▮▮▮▮▮▮▮▮▮▮▮▮▮▮▮▮▮

MIZLING ALONG

izle is a word I've never spoken out loud, nor even
written before this moment, although it's been
nestling in the niches of my mind for about ten years, and
there have been frequent moments when I've hauled it
out and savored it silently. It was about ten years ago that
I first heard a friend use the word. He and I were writing
a story that we hoped would be a good screenplay. Unfortunately, our interest flagged just as, I fear, an audience's
interest might have, approximately halfway into the
catastasis.

But in the course of one afternoon, Bill must have said
mizle, mizled, or *mizling* five or six times. That's pronounced with a long *i*, of course: *my-zl*. The first few times
he said it, I didn't know precisely what he meant, but I
didn't say anything, and I'm glad now that I didn't. In context, it was clear that to mizle was not a nice thing. It had
a kinship with *bamboozle, bilk,* and *swindle*, I felt.

After a time, I thought I perceived what had happened,
and a total picture was revealed to me. At some moment
in his past, Bill must have come across the past tense of

mislead. He had misread *misled.* In his mind, *misled* had been the past tense of *misle,* and it meant whatever he thought it meant. It was connected tenuously to *mislead,* but it also meant swindle, gyp, double-cross, and shaft. It has a right sound for those meanings. It sounds mean.

I never discussed the word with him. I just let it ride, so my explanation of *misle* is purely conjectural; still, I feel sure it must be accurate. Perhaps I should have mentioned it. I think I may have been hesitant to embarrass him, but I know too, that I liked the word and I wanted to hang on to it in all its mystery. In my mind, I've used it often, but when I see it internally, it's always with a *z,* to differentiate it from the past tense of *mislead.* The only other words I can think of that have *isle* in them (other than *Paisley,* which is not really germane), are *isle, aisle,* and *lisle.* With that in mind, Bill just might have read *misled* as *mild.* Now we're in that old familiar field: The Weird Orthography of English.

Trying to think of other words that might be subject to this sort of misleading misreading, I've come up with only one, though there must be many. There's a remote possibility that the first time you saw the word *pitiable,* you'd lose sight of *pity.* If so, your inner ear would probably register *pishable,* the *ti* being *sh* as in *negotiable* and the classic *ghoti* for *fish.* Thence you'd back-form a verb *pitiate* (rhyming with *vitiate*).

Misle from *misled* and *pitiate* from *pitiable* are reasonable expectations, given the normal structure of English words.

I've been intrigued by comic-strip characters whose lack of intelligence and/or education is indicated by their pronunciation. The first one to impress me with this syndrome, way back in my hors d'oeuvre days (that carefree period preceding the salad days) was Joe Palooka, who said *lissen* for *listen, peeple* for *people, sumthing* for *something,* and so on. Ham Fisher, Joe Palooka's creator, was doubtless indicating that his hero, though golden-hearted, was leaden-headed. Somehow, it never worked—for me, anyway. A guy who said *lissen* for *listen* was okay with me; he knew just what he was doing. I never fancied that Joe Palooka wrote his own stuff. A comic-strip character can't help what his writer gives him; he just says it out.

I mention Joe Palooka only because he was the one who set me thinking about this grave question so very long ago. There are dozens of other cartoon characters who speak equally well, while their writers are doing an abominable job of spelling. If the cartoonist misspells *listen,* that's not the cartoon character's fault.

I guess if Joe Palooka had ever used as highfalutin a word as *misled,* he'd have said *miss-led,* the big lug.

Personally, I don't think I'll ever use *mizled* again, even to myself, now that I've brought it out into the open. And I hope *nobody ever* uses *pitiate.*

March, 13, 1973

SHOPPES 'N' THINGS

There are a couple of things about the name Gourmet Health Food Shoppe that jar my mind. I just saw those four words on the awning of an establishment. The three concepts are wrong together.

The word *gourmet* doesn't belong with *health food*. I can hear the roars of outrage from an army of alfalfa-sprout *aficionados,* and I sympathize and plead that I am speaking very strictly. A true gourmet is concerned with the sensual pleasures of the table and doesn't care a marinated fig for the effects of food and drink upon his physical well-being, short of an Amanita omelet.

In our kitchen, Adelle Davis is between Julia Child and *The Gourmet Cookbook.* We use raw sugar mostly, but not exclusively, and our favorite bread is so natural it has rocks in it. We use cold-pressed soy oil and sea salt, but we also use olive oil and Morton's. In other words, I don't mean that "health food" can't satisfy the most exacting gourmet; I'm saying that a true gourmet, qua gourmet, doesn't ask the question, "Is it healthful?"

But the stunner, more than "Gourmet Health Food," is "Shoppe": Gourmet Health Food *Shoppe?*

Shoppe goes with *Ye Olde*. It's from Middle English and hasn't been used since about 1800, except as an affectation on tearooms and antique shops.

The first use of the word *gourmet* in English noted by the *Oxford English Dictionary* is in 1820. "Gourmet Shoppe" is an anachronism as well as an abomination.

Personally, I'd like to see *shoppes* die out altogether. Back in 1841, Dickens had better taste than to call his novel *Ye Olde Curiosity Shoppe*. Now, 132 years later, *shoppe* is ludicrous.

Another kind of shop I'd like to see die out is the " 'n' Things" shop. I don't know about America's other major cities, but in New York and Los Angeles, a plague of " 'n' Things" shops arrived about ten years ago. (Or was it twenty, or was it five? Time seems increasingly to lose its definition as fads and fashions come and go in greater profusion and with greater rapidity.) There was a sudden outcropping of little places called Rings 'n' Things, Lamps 'n' Things, Sandals 'n' Things, and so on *ad nauseam*. They appear to be losing ground lately. I can see why. No self-respecting adult should be expected to walk into an " 'n' Things" shop. He'd as soon walk into an Adult Bookshop (what an astonishing appellation for a porno den). The " 'n' Things" shops seem to be going out of business, which reminds me of Stanley Myron Handelman's brilliant concept of the rich kid who inherited a chain of Going-out-of-Business Stores.

Getting down to basics, I'd like to see *'n'* go out of business. I won't buy a can labeled Pork 'n' Beans. I won't even order Ham 'n' Eggs when that's how it's presented on a menu. A friend of mine once said he'd like to publish his own dictionary just so he could call it *Words 'n' Definitions.*

It's true that hardly anyone pronounces *and* with all its values. No one can say, "Honorable and just," sounding the *d* before the *j*, without sounding a bit phony, except someone with the meticulous speech of Sir Alec Guinness, and he'd even pronounce the two *s*'s in *meticulous speech,* with the slightest break between the *s*'s. It takes class to do that and get away with it. But *and* is a most important little word, and *'n'* is an insult to it; *'n* and *n'* with only one apostrophe, are even worse.

At least one instance comes to mind in which shabby treatment of *and* has caused linguistic confusion. "The spit of" is a time-honored phrase for "the exact likeness of." "The spit and image of" was an elaboration of the phrase. It's colorful. "She's the spit and image of her grandmother" has a sort of swagger to it—a pleasant style. Unfortunately, *and* was too often pronounced *'n',* so that the phrase was tortured to "the spittin' image of her grandmother"—an entirely different and less pleasant idiom. Neither phrase is elegant, but "the spit and image" is likable. "The spittin' image" is a slob.

I confess that, for love of Burnett *'n'* Matthau, I saw *'n'* enjoyed *Pete 'n' Tillie,* but I wouldn't take annie oakleys to see *Romeo 'n' Juliet* or *Tristan 'n' Isolde.*

March 27, 1973

WEASEL TO HERRICK
TO STRAUSS

The *Light Refractions* about "Pop Goes the Weasel" elicited an unpredictably large response. I revealed the truth about W. R. "Handy" Mandale, his tailor, Harry Krishner, and their part in the creation of what must sound to the unenlightened like a nonsense song.

Leonard Marcus, editor-in-chief of *High Fidelity*, wrote in to cite *The Book of World-Famous Music*, a scholarly work by James J. Fuld. I wasn't familiar with the book, but felt I should probably own it, so I bought a copy. None of what Mr. Fuld says about "Pop Goes the Weasel" jibes with my version of the story. I won't go into detail, except to say that in his researches, he evidently found nothing about "Handy" Mandale or his drunken tailor.

Mr. Luigi Rothchild wrote to tell me that Eric Partridge's *Dictionary of Slang and Unconventional English* has a paragraph on "Pop Goes the Weasel." It wouldn't have occurred to me to look there, but sure enough, Partridge says that "Pop goes the weasel" was a proletarian, mostly cockney, catch phrase, and he quotes

J. Redding Ware's *Passing English* as saying it's probably erotic in origin. Make of that what you will.

My favorite piece of correspondence was a cartoon from *The Christian Science Monitor* sent by Mr. Ernie Scheuer. The cartoon is by H. Martin, and it's in three panels. The first shows a couple watching a small animal in a field. The man says, "That's your opinion, Mildred. I say it's a ferret." In the second panel, the little animal disappears with a "pop," and in the third panel, the man says, "You're right. It was a weasel."

I'm grateful to Mr. Marcus for alerting me to Mr. Fuld's *Book of World-Famous Music*. It's a remarkably catholic work—folk songs, hymns, jazz, Bach, Brahms, Beethoven, Burt Bacharach: a huge spectrum, the sort of book that's fun to forage in. He devotes almost six pages to "The Star-Spangled Banner." I suppose it's common knowledge that the tune of "The Star-Spangled Banner" is taken from an old English drinking song, but it took Mr. Fuld to tell me more about it. The drinking song was called "The Anacreontic Song," and its opening line was, "To Anacreon in Heav'n, where he sat in full glee." That piqued my curiosity.

I didn't now who Anacreon was, so I went to the encyclopedia. He was a Greek poet born in the sixth century B.C. His poetry was mostly about wine and love. Very little of his work survives, but his reputation was kept alive through centuries of imitators who wrote Anacreontics— poems supposedly in the style of Anacreon.

At the library I found a couple of books of translations of Anacreontics, generally on the order of "Bring the Wine Bowl, Ganymede" and "Young Venus on a Mossy Bank." I can't take too much of that at one sitting.

In the seventeenth and eighteenth centuries some of the better English poets took up the writing of Anacreontics, among them Thomas Gray, Matthew Prior, and Robert Herrick, and a lawyer named Ralph Tomlinson wrote the words to "The Anacreontic Song" for the Anacreontic Society, a London social club specializing in singing and conviviality. No one knows for sure who wrote the music, but it became a very popular tune in both England and America. It was used for about eighty-five different printed American poems from 1780 to 1820, according to Fuld.

"The Star-Spangled Banner" was again an object of controversy a few months ago because of the way it was sung at the presidential inauguration. This sort of thing is happening more and more frequently these days, for obvious reasons. If you want "The Star-Spangled Banner" sung straight, it seems to me that you should ask a singer of the traditional school to sing it. If you ask a popular song stylist to sing it you shouldn't be surprised at a non-traditional rendition. I admire anyone who can get through it at all.

I read one letter to the editor on the subject that bewailed the fact that no one was singing "The Star-Spangled Banner" in the good old quick-march tempo any more.

I'd never thought of "The Star-Spangled Banner" as a march, so I was surprised to read in the *World Book Encyclopedia* that the tune had been a popular military march in the 1700s. It's in 3/4 time. In the 1700s was cadence counted differently? Instead of "Hup, hoop, heep, hawp; hup, hoop heep, hawp!" it may have been "Hup, hoop, heep; hup, hoop, heep!" and soldiers may literally have waltzed into battle.

Back in my marching days, cadence was often hollered, "LEF'! LEF'! Lef' ri' LEF'!" Try doing that to "The Star-Spangled Banner." It comes out, "LEFT right left, RIGHT left right, LEFT right left, RIGHT left right." Makes you think of Johann Strauss.

See where a little curiosity about a weasel can lead?

April 10, 1973

SUPERGHOST

A couple of weeks ago I mentioned that people who can't spell, spell death to a game of superghost. Since writing that, I've brought up superghost in conversations with friends and have been astonished to find that there are people who've never heard of it. I hadn't realized, either, what a long time it's been since I last played a real live wide-awake game of superghost. I assume that most readers of *Light Refractions* are logophiles and are familiar with this word game of word games, but for you who are not, I'd better describe it.

Ghost is a game in which one person starts a word by saying one letter, and the other players in turn add one letter each until a word is spelled. The point is to avoid being the one to speak the final letter that forms a complete word. Finishing a word makes you a *g;* finishing two words makes you a *g-h,* and so on until you're a *g-h-o-s-t.* Then you're out of the game.

In superghost you can add not only to the end but to the beginning of a group of letters.

I remember years ago reading a piece by James Thurber

based on superghost. I'm pretty sure I remember the title of his piece. It was brilliant: "If You Put an *O* on My Understo, You'll Ruin My Thunderstorm." That tells you pretty much what superghost is all about.

Thurber said that when he found that he couldn't sleep in the middle of the night, he would think of letter combinations for imaginary superghost games. I experienced that almost smug, warm identification with the great, knowing that Thurber and I did the same thing while suffering from insomnia.

I actually have a cast of characters with whom I play superghost. One of them—the slight, balding one with glasses who's sipping a martini—starts with a *j*. The shaggy Matterhorn of a man with the curly black hair who affects a beard, an English accent, and a black turtleneck sweater, makes it *d-j*. The next player is the wise guy—a know-it-all who chews a cigar and speaks in a slow, snarly nasal drawl in imitation of Ned Sparks. "*D-j-a*," he says, killing the possibility of *adjutant, adjective, adjusted,* and *maladjusted,* among others. With *d-j-a*, it looks like *adjacent* or *non-adjacent* (an awkward word at best) or variations on them, *nonadjacency, adjacencies,* and so on. It's my turn. "*D-j-a-m*," I say. My opponents are puzzled. I can see I have them on the ropes already. Big English Accent tries to joke away his consternation. "There's no *d* in *pajama,* old boy," he says. "He's bluffing," says the martini sipper. "I challenge you." "*Windjammer,*" I say.

The gratifying thing is that while my fellow players are all very clever, they can't win. Notice how none of them thought of *windjammer*. The other night we had arrived at the combination *u-t-h-a*. The obvious word to grow out of *u-t-h-a* is *euthanasia,* I thought. How could I stun them and stump them? First I thought of *u-t-h-a-t*, going for *nuthatch*, but finally settled on *u-t-h-a-u*. Imagine the gnashing of teeth when, being challenged, I said, "Outhaul."

The fun of the game, of course, is in discovering unusual letter combinations and developing them in mysterious directions. It calls for seeking the unobvious. If your combination is *g-h-t-f*, the first words that come to mind are probably *thoughtful, rightful, delightful,* and other *ful*-suffixed words, so you scratch around a bit and find *nightfall, lightfoot,* and *tightfisted. Tightfisted* will cause trouble, because opinion will be divided as to whether it's hyphenated or not. I wouldn't leave my bed on a chilly night to get a dictionary's decision to satisfy my imaginary superghost companions, and in my experience most real superghost games are played in bars or automobiles, neither of which is apt to have an available dictionary. So in the best superghost games, hyphenated words are not allowed, though I was in a game many years ago where they were allowed, but you had to call both the hyphen and the letter on the other side of it, and one player whose name I can't recall did a stunning job with *t-c-h*. He said, "*T-c-h-* hyphen *c*," was challenged, and brought out *catch-as-catch-can,* which was greatly admired. Generally, though, hyphens cause more arguments than they're worth. Someone's bound to pull something like *leather-hearted*; someone else is bound to shout that that's ridiculous; and trouble is bound to result, especially if the game is taking place in a bar.

A salutary side effect of nocturnal, solitaire superghost is that it becomes an effective soporific. As the mind gets more and more letter befogged, weird combinations emerge. I dimly remember falling asleep in the early morning hours while trying to get from *t-h-r-a* to the word *thraight*. "I've forgotten how to spell *thraight*," I thought. "Is it *thraight* or *thrait*?" Just as I dozed off, I realized that *thraight* meant "straight through," and must be spelt *t-h-r-a-i-g-h-t*. When you reach that state, sleep is at hand.

May 8, 1973

OBSCENITIES, OLD AND NEW

When I saw *Mr. Roberts* on the New York stage, I was with my parents and my younger sister. I think it was Harvey Lembeck who, using binoculars to inspect the interior of the nurses' shower room, roared, "She's got a boit'-mark on 'er ass!" My father confided in me that he was sorry we'd brought my sister to such a racy play. At the time, she was twenty years old.

I was reminded of that a few weeks ago while we were visiting our son at the University of California at Berkeley. He's majoring in the conservation of natural resources.

Tom lives in a three-story Victorian frame house with seven other young men and women.

They call their living arrangement an "ecological habitat." They live their field of study. The only meat they eat is meat that they grow and slaughter—rabbits and chickens. They utilize waste products to feed their animals and to make compost for their approximately one acre of farmland. They grow worms in trays of rabbit manure and feed the worms to their chickens. Practically nothing is wasted in their ecological system.

There's a large receptacle in the kitchen and a sign that says:

COMPOST

This bucket is not a garbage can—it is a compost can. Compost is a process used to change organic wastes into soil. It is something that nature does naturally like the way we breathe or shit. There is a whole host of characters that help turn organic matter into soil. They are bacteria, fungi, worms, insects, and other decomposers.

Organic matter breaks down at different speeds. The speed depends on how easy it is for the organisms to "chew" the material. Smaller pieces are easier to digest, so please cut everything into small pieces for our little friends. Plastic, wood, metal, rubber, cloth, and similar things have other ways of being reused. Only vegetable matter goes into our compost. To keep down the odor we sprinkle a layer of sawdust on top before we go to sleep at night. To us good compost is worth more than gold. Ask if you want to know more about our compost or recycling or anything else, and turn us on to what you know or feel about such matters.

Not only are you struck by the startlingly cavalier attitude toward gold, but if you're of my generation or older, your eye was arrested and perhaps somewhat offended by that little four-letter word.

Twenty-five years ago, my father was disconcerted by the line in *Mr. Roberts*. I confess to having been equally taken aback by the use of the four-letter word instead of the more acceptable three-syllable one.

Tom showed us where they'd cooked the rabbit we'd eaten the night before. They have a tank shaped much like the tank on an oil truck, but only about six feet long.

It contains chicken and rabbit droppings, which produce methane gas. Tommy opened a valve, struck a match, and had a clear blue methane flame. In describing it to us, he didn't say "droppings," of course. He used the more vulgar term.

It's possible to consider the very casual use of words labeled "vulgar" in our dictionaries as an outgrowth of the rather silly "dirty words" free-speech campaign that erupted on the Berkeley campus a few years ago, but I think there's more significance than this, and I don't think it's a bit of youthful rebellion or affectation.

It's not unlikely that the reversion to the ancient four-letter words is related to a keen awareness that we've got to live in a more elemental relationship to our natural functions, an awareness Tom and his colleagues refer to as "turd-consciousness." Much of what we've chosen not to think about, much of what we utterly reject as "indelicate," may, it seems, ultimately prove to be of vital importance to our survival if we can manage it wisely. So the terms are not offensive to those who know the value of what the terms refer to. Symbolically, there is a value in eschewing euphemisms when you're dealing positively with what has heretofore been dealt with only by removing it as far as possible from the minds of nice people.

I suspect that there must be taboo words in any great language. Civilized society needs its unspeakables. The words that were taboo to us will probably become acceptable to later generations. What I see—or am I wishfully thinking?—is a new body of taboo words for the language. The new taboos are words that insult, words that denigrate people in terms of race, nationality, sex, and so on—the body of nasty words for people who are in some way different from the user of them. That, I hope and believe, will be the new obscenity.

It's worth thinking about.

June 5, 1973

PLEONASMS AND REDUNDANCIES

On May 23, Sen. Hugh Scott was asked what he thought of the possibility of the President's being impeached as a result of the skulduggery at the Watergate. Senator Scott replied, "We don't deal in the incredible and the unbelievable."

When I heard this reported on the radio, a little flag labeled Redundancy sprang up in my mind with pedantic precision. The Redundancy flag didn't involve any sense of irritation. This particular redundancy is so common, at least in my experience, that far from holding it as anything like a pet peeve, I've come to regard it as an old friend. Usually (though not, I think, in Senator Scott's case) it's used in attempting to express something so out-of-the-ordinary that words fail, while loquacity doesn't: "What a backhand! It was really incredible! It was so incredible you could hardly believe it!" Hyperbole is the norm these days, as in, "How's the hamburger?" "Superb!"

When I hear "It's so incredible, it's unbelievable!" I have a little trouble believing how incredible it is, and I let it go at that.

I have some favorite redundancies. I think it was in *The Count of Monte Cristo* that Robert Donat's final line, spoken directly to the camera, was something about how the enemies of Monte Cristo had been brought low by those two fatal flaws in their characters—dramatic pause— "Avarice . . . and Greed!" Identical twins, you might say.

I used to like the way Wyatt Earp, as portrayed on television by Hugh O'Brien, was glorified as "brave, courageous, and bold" in the show's theme song. The song didn't mention it, but I guess we all noticed that he was also valiant, intrepid, and plucky.

The redundancy of the Wyatt Earp song was the result of a need for syllables to go with a certain number of beats and a lack of inventiveness on the part of the lyricist, but Senator Scott's redundancy and Monte Cristo's redundancy are both caused by a natural liking for couples. I see the same phenomenon in my daily life. A typical domestic conversation might go like this:

> *Self:* I'm going to the store.
> *Wife:* Good. We need two things.
> *Self:* What?
> *Wife:* Coffee.
> *Self:* What else?
> *Wife:* I can't remember.

The fact is, of course, that we needed only one thing, but it sounds better to say "We need two things," just as Senator Scott must have felt there were two things he didn't deal in, and Monte Cristo felt his enemies had two fatal flaws. One is just too lonely.

I think I must have been about nine or ten years old and already an inveterate label-and-cereal-box reader when I was deeply impressed by what used to be the Squibb Company's slogan: "The priceless ingredient of every product is the honor and integrity of its maker." I must have read that on either a toothpaste tube or a bottle of milk of magnesia. All I know for sure is that I was standing at the washbasin in the bathroom and I thought this saying was one of the most profound and beautiful sentiments I'd ever encountered. I whispered it to myself several times until I'd committed it to memory: "The priceless ingredient of every product is the honor and integrity of its maker."

It was several years before it occurred to me that, in this sentence at least, *honor* and *integrity* mean the same thing. Each word has several meanings, of course, but what is clearly intended here for both is "probity, uprightness, and adherence to moral principles." *Honor* and *integrity:* words that are brave, courageous, and bold. Obviously, the point is that in many cases a little redundancy is not such a bad thing. That slogan has stuck in my mind for all these decades, unforgettable as "Don't give up the ship," "Don't fire until you see the whites of their eyes," "Remember the Alamo," and "Don't tread on me." If they'd left out either *honor* or *integrity,* they'd have avoided the redundancy, but it wouldn't have had the august ring that so impressed my young mind.

Suppose, at the end of *The Count of Monte Cristo,* Robert Donat had wrapped things up by enumerating his enemies' *one* flaw: "Greed." What kind of dead-end enumeration would that have been? It would have been another of the cinema's millions of forgettable lines.

If, when Senator Scott was asked about the possibility of impeachment, he'd replied simply, "We don't deal in the incredible," it wouldn't have triggered these thoughts; it would merely have established the senator's incredibility gap. It would have been remarkable only as an unbelievable indication that Senator Scott considers impeachment incredible.

I can hardly believe it.

July 17, 1973

AUTHORITIES' TALK UPDATED

The Watergate hearings have yielded a stunning display of Authorities' Talk, that sterile speech in which *people* are *individuals, cars* are *vehicles, no* is *negative,* and so on. Former CIA men seem to talk that way. I wonder if active CIA men do, too. It seems to me that anyone who works as a spy would profit by learning to talk like an ordinary, everyday person, just to keep from blowing his cover.

In the training of a CIA operative, there should be courses in the casual, unpretentious use of the English language. Perhaps there are such courses, but it now seems more likely that we are to be so bombarded with Authorities' Talk that we'll come to consider it quite normal. We've heard otherwise articulate news commentators lapse into it lately.

Listening to the early Watergate witnesses, I began to think "then" and "now" had become archaic. They were superseded by "at that time" and "at this time," which in turn have been superseded, at least in the Watergate hearings, by "at this [or that] point in time."

"At that point in time" has the virtue of sounding precise. Time, as the fourth dimension, is arrested very specifically at a point—an infinitely fine mini-moment during which an undiscovered number of angels may or may not be dancing in 0/0 time. The trouble is that "at that point in time" is used in virtually every instance to refer to a vague period (not point) of anywhere from a few days to several months. So whatever virtue "at that point of time" has over "then" is vitiated.

Jeb Stuart Magruder, in fact, referred to a "point in time" when there was active, vocal, and sometimes violent opposition to the Vietnam war. That would cover two full presidential terms, at the very least. Some point!

"At that time" and "at that point" are both useful phrases, of course, as in "*At that time,* I was working for Wetsy-Betsy Doll Company. I was a bugger on their breaking-and-entering team connected to their inter-enterprise espionage division. As part of our game plan, we broke into the Killer Kommando Fun Toys building. We were apprehended by an individual whom I surmised was a security guard, and *at that point,* I foresaw implications that the operation, which I had considered a legitimate function of the free enterprise system, might be criticized by a hostile judiciary."

As a substitute for either of the italicized phrases, "at that point in time" would be completely inappropriate, but that wouldn't deter anyone properly versed in Authorities' Talk.

I'm sure everyone's noticed that, in addition to its several traditional meanings, savory and unsavory, *bugger* has acquired a new one: one who engages in bugging. Not very savory.

"Watergate" may well become a verb. I remember, in 1951, hearing a young lady who was being badgered by her boy friend on the question of her recent whereabouts. Suddenly, she'd had more than enough, and in angry retaliation, she cried, "Don't Kefauver me, dammit!"

So, "to Kefauver" became a verb, however briefly. "To Watergate" will probably have at least a flash-in-the-pan existence. It's not hard to imagine, a few years from now, reading in the paper that a group of wiremen have been caught in the act of Watergating some office or other.

You probably noticed that my Wetsy-Betsy Watergater said "an individual whom I surmised was a security guard." That's standard Authorities' Talk. The improper use of "whom," as in "an individual whom I thought didn't belong in that particular area," is as predictable as the use of the subjective pronoun in any and all cases when two people—pardon me, *individuals*—are mentioned: "We took he and his wife to dinner that night"; "the alleged suspect accosted Officer Corrigan and I"; etc.

There's also the all-purpose first person singular pronoun: myself. Anyone who's not quite sure whether to use "I" or "me" assumes that "myself" will cover any situation. Being neither subjective nor objective, it will serve for both, presumably. "Myself and my partner descended from the vehicle," or "he was looking in the general direction of myself." This latter is a doozy. The speaker has just muffed the opportunity to speak simply ("he was looking in my direction"), and finding himself precariously close to saying, "he was looking in the general direction of me," which is almost unutterable, he grabs the nearest substitute, the serviceable *en tout cas*, "myself."

That's the real trouble with Authorities' Talk. In avoiding the simple ways of saying things (for fear of sounding simple? or of sounding human?), its practitioners come out sounding pompous and, I'm afraid, not very bright.

July 31, 1973

THE DIVINE DIZZINESS

To me, the screwball uses of language are among life's great joys. I think Peter De Vries has the keenest ear in the English-speaking world of linguistic absurdities.

One of my cousins, who lived, he said, in RATchester, New York, pronounced "probably" as "prally" when he was in his early teens. Years later I met a young lady from WisCANsin who pronounced it "parry," as in "Tamarra'll parry be even hatter'n taday." But it took Peter De Vries to find a young lady who said "pry." I loved her instantly. I can't remember her name, but I'm sure she was in *Reuben, Reuben,* one of the many De Vries books I've lent or given to friends in the course of proselyting for the De Vries cult. I pry lent it to a friend who pry lent it to someone else.

In De Vries's latest book, *Forever Panting,* he has created a woman who talks like my wife.

The De Vries character is named Ginger, and she talks in Ginger-isms. For years I've marveled at and loved what I've called Jeannie-isms. "She was one of those women who manage to be both dizzy and intelligent," says De Vries of Ginger.

This is the sort of sentence you can get away with in a novel, but if I referred to a woman, especially my wife, as "both dizzy and intelligent" today, hordes of modern women would bridle justifiably in recognition of the stereo-typical "dizzy dame." Nevertheless, it is a fact that there are some women who have the divine gift of natural, unintentional comedy, and Jeannie is among them.

Some men have it, too, of course. The most famous Dizzy is one of the Dean boys, who is as famous for saying "He slud inta third" as he is for his phenomenal pitching. It was Joe Jacobs who said, "I shoulda stood in bed," and Sam Goldwyn, who was anything but dumb, made—or at least was credited with having made—hundreds of dizzy statements, my favorite of which is, "A verbal contract isn't worth the paper it's printed on."

Back to Ginger-isms and Jeannie-isms. "The scatter-brained woman is not the worst company in the world, if it's brains she's scattering," says De Vries through his leading character, Smackenfelt. Far far, far from the worst company, I add from personal experience.

Ginger says things like, "People in show business always have a lot of irons in the fire, but few of them ever jell."

Jeannie says things like, "If I had the time to play bridge, I'd take up all my hems."

One of the classic Jeannie-isms was: "Gee, I was really surprised to hear about Lee and Betty getting a divorce. It didn't surprise me, though."

I was reminded of that a couple of weeks ago when I got a card from a lady whom I'll call Mrs. Tyler. Critical mail seldom delights me, but Mrs. Tyler's card made me feel right at home.

> I am totally amazed that you—a self-proclaimed master of linguists
> and a crusading purist for the English language—can so befoul
> the same by your misuse of "elemental" [in an earlier article not
> reprinted here]. I am, however, not surprised—as week after
> week you submit inaccurate "clues" to the puzzle.

What can I say? I'm totally disarmed by anyone who's
"totally amazed" but "not surprised." Mrs. Tyler was referring to something I said about George Orwell's Newspeak in his *1984*. I said that, in Newspeak, "verbal thought
becomes possible only on the most elemental level." I
meant elemental in the sense of "not complex or refined"
(WNI 3) and "starkly simple, primitive, or basic" (Random House).

Mrs. Tyler must have been thinking of elemental as
"relating to or caused by a great force of nature." That,
indeed, would have been a misuse of elemental.

Being totally amazed but not surprised seems to be
much more a feminine than a masculine aptitude. In fact,
in a poll of more than a half-dozen women, I found only
one who was in the least surprised that Mrs. Tyler was
totally amazed but not surprised, and even *she* wasn't
surprised that Jeannie was really surprised but that it
didn't surprise her, though. On the other hand, it's been
my observation that men are much more prone than
women to use the word "literally" to mean its exact opposite, sometimes evoking bizarre images: "He was literally
coming part." The other day I heard a state senator say,
"I had Senate bills literally coming out my ears."

That fractured me. Literally.

September 11, 1973

BUNKINTA

I got a very nice letter from S. J. Tress, "a teacher of English in Brooklyn." Tress didn't specify whether he or she is Mr., Mrs., Miss, or Ms. That's not what I want to write about, but Mr., Mrs., Miss, or Ms. Tress did bring up something that set my mental juices flowing happily, and I want to give him or her credit. One of today's gravest problems is whether or not to refer to a woman as "him" or even as a part of "mankind" and risk being a sexist pig or whether to assume that "a teacher of English" is a woman and risk being a sexist pig if she *is* a woman or a mentally retarded bigot if he is a man. Anyway, I'm grateful to S. J. Tress for the letter, which closes with, "Did you know that New Yorkers use the word *bunked* instead of *bumped?* They say, 'I *bunked* into my old friend on the avenue.'"

I had long been aware of *bunk* for *bump*. For many years after moving to Los Angeles from New York, I still thought of myself as a New Yorker. That's a common phenomenon. I found New York speech patterns fascinating and still do. As I remember "bunked into," though, it was one word:

bunkinta. "Ya know who I bunkinta inna fawdy seng streedawdamat?"

The speech of New York City and its environs deserves volumes. Discounting the many foreign and ethnic patterns, there seem to me to be two basic dialect divisions that I call the New York accent and the New Yorker accent. The extreme New York accent is often referred to —imprecisely—as Brooklynese. As a convenience, I'll go along with that for now. The extreme New Yorker accent has been called the High Episcopalian Throat, or Hoy Episcopalian Threywt, as its practitioners say.

Louis Nye did a very creditable High Episcopalian Throat in his character Gordon Hathaway, who said, "Hoy heyw, Steverineyw!" on the old Steve Allen TV show. The High Episcopalian Throat is best accompanied by almost unbearable languor.

The fact is, of course, that there are many New York accents—fascinating divisions, subdivisions, and blends of subdivisions. Come to think of it, Mr., Mrs., Miss, or Ms. Tress is right about "bunked into." New Yorkers who do not say "fawdy seng streedawdamat" often *do* say "bunked into," not "bunkinta." "Bunked into" sounds right. "Bunk" is at least as onomatopoeic as "bump."

As I've mentioned before, accents are changing and becoming homogenized. The extreme Brooklynese *r* used to be pronounced almost as a *v*. I learned that in the army. As any historian of World War II knows, every military outfit had a Kid from Brooklyn, and ours was no exception. Tony McAloney (that rhymes with Baccaloni) was our Kid from Brooklyn—a young man with a brilliant, inquiring mind, a love of weight lifting, and great skill in boxing. It was Tony who suggested that the Brooklyn *r* was said as *v*. We tried it. Throughout the shooting of a pool game, we consciously substituted *v* for *r*, as in "Vockefella's a vevvy vich man." And no one noticed anything strange; it was so close to what was then standard "Brooklynese." Thanks largely to the influence of Huntley, Brinkley, Cronkite, Robert Young, etc., that's changed. Somewhere there may still be someone who says t'oidy t'oid stveet an' t'oid avna, but it's no longer commonplace in New York, as it once was. There are still vestigial traces, but the inroads made by standardization are unmistakable. I think it's a shame that dialects are being smoothed out of existence, but they are.

Here in Los Angeles, the interesting accents are all from other regions. A couple of peculiarities arise in the pronunciation of *Santa Ana* and *Los Angeles.* In Southern California Santa Ana can refer to a town, a freeway, or a local hot wind. Whichever it is, it's almost always pronounced Sannyanna.

In *Los Angeles* the *Los* is pronounced variously to rhyme with gross, cross, and joss, and the *Angeles* goes Angle-ess, Angle-eez, An-jeless, and An-jeleez. Sam Yorty, our former mayor, simplified it. He called Los Angeles "Sanless."

In L.A. (the one standard pronunciation for the city) we may say "bump" instead of "bunk," but in New York if you bunk into your old friend on the avenue, it's nice. Out here, if you bump into a friend on the avenue, it will probably involve auto insurance companies and lots of repair bills.

But don't get me wrong. If I didn't like it here, I'd go back where I came from.

October 9, 1973

WHAT'S THE SQUARE ROOT OF SUPER COLOSSAL?

The other day my wife asked me to pick up some large ripe olives at the supermarket. Easy enough. I wandered around, looking at the little signs that hang over the aisles until I saw one that said Pickles Olives Relishes. A shelf in the olive department had a selection of a brand of California ripe olives. One size was Extra Large. That should do, I thought, but let's see what the just plain Large olives look like. No Large. In a very short time, I learned that not only was there no can of California ripe olives labeled simply Large but also that Extra Large was the smallest they had. The other choices were Giant and Super Colossal. This may, of course, have something to do with the fact that they're *California* ripe olives. Maybe the glamour of supercolossal, stupendous, gargantuan Hollywood films rubbed off on the California olive trade a few decades ago, and they're stuck with it. Habits like that are hard to break. But are they thinking big or thinking small? It seems to me that nothing smaller than Roosevelt Grier should be called Colossal, and surely nothing Super Colossal should be bite-sized.

Another example of peculiar size labeling can be found in the beer section of your market. The standard beer bottle used to contain twelve fluid ounces; and when beer started coming in cans, *they* all contained twelve fluid ounces.

Well, as everyone nows, a few years ago some of the big brains of the beer merchandising world came up with the brilliant idea of putting out sixteen-ounce beer cans. Problem: What should they be called? If your answer is "pints of beer," it just shows how simple-minded you and I are compared with the heavy thinkers of the beer-selling industry. I've come into possession of some tapes of one of their meetings. Here's an excerpt:

> "Brilliant idea, Maurice! Sixteen-ounce cans! We'll call it our new Pint Size!"
>
> "Pint Size? No, no. It doesn't quite sing to me. There's something about it that doesn't set right, and I think I know what it is. Pint Size sounds dinky."
>
> "You're right! Pint is small. Quart is big. You might as well say Half Pint Size as Pint Size."
>
> "I think you've put your finger on it, Harry. Quart is big. Pint Size means the same as Half Pint—small. How about Half Quart Size?"

(At this point on the tape there is a general brouhaha, and most of the words are unintelligible. However, there is a sense of jubilation. Finally, one of the speakers comes through clearly.)

> "Maurice, you're a genius! Our new Full Half Quart Size! Now *that* sings!"

So in the United States today, there's no such thing as a pint of beer, but there are millions of Full Half Quarts.

When you think about it, that's millions of Double Half-Pints.

It's mind boggling, and I suppose that's the point.

About fifteen years ago, almost every brewery in the West had an equally brilliant, though completely different, concept in beer packaging. What they did was to subtract one from twelve and put the newer, more compact eleven-ounce can on the market. They did it with no fanfare whatsoever. No big ads saying, "New!!! 8⅓% less for your money!! 66 fl. oz. per six-pack instead of the cumbersome old 72 fl. oz.!!!"

They just sold the smaller cans without saying anything about it.

I don't know what made them put the missing ounce back and revert to the old twelve-ounce cans, but I think they all did so after a couple years. I hope the deciding factor was that enough people noticed that the cans said "11 fl. oz." instead of the familiar "12 fl. oz." and that, like me, they took barrels of umbrage and stopped buying the beers that were guilty of this quiet bamboozlement.

Let's face it, who needs 11/16 of a Full Half Quart for the price of 3/4 of a Full Half Quart?

October 23, 1973

ENTY?

Mr. George Stevens was the managing editor of *The Saturday Review of Literature* when Elizabeth Kingsley first brought her puzzles to the magazine back in the early Thirties. It was he who thought of the name Double-Crostics.

He wrote me to say, among other things, "I do not know how many words there are in the English language, but in addition to the hundreds of thousands that exist, two more are needed.

"First, an adjective for *would-be*—as in 'a would-be poet,' 'a would-be musician,' etc. It strikes me as particularly awkward that we cannot do better than to translate *Le Bourgeois Gentilhomme* as 'The Would-be Gentleman.'

"Second, an equivalent for '*n'est-ce pas?*' covering all persons and both numbers, to avoid the pedantic awkwardness of 'am I not?' 'isn't he?' or even 'aren't you?' Actually, there is such a word, but it lies neglected, even by the unabridged *Random House Dictionary*: the word *enty*. I have seen *enty* in print in only one place, *The Complete Tales of Uncle Remus,* and it is not Uncle Remus who uses it but Daddy Jack, who speaks Gullah. The only other person I ever knew who was acquainted with the word was the late Clyde Brion Davis, author of a long string of novels beginning with *The Anointed*. This is the novel that was made into a movie under another title—I forget what—and was advertised directly after the Second World War with the irresistible line, 'Gable's back, and Garson's got him.' I don't know where Clyde picked up *enty*—he never lived in the South, and I doubt if he ever read *Uncle Remus*."

I don't share Mr. Stevens's uneasiness with *would-be*, but *enty* wins my admiration. I'm in favor of its inclusion in future dictionaries. It's a beautiful, simple word, and it fills a need that I, too, have felt for an English equivalent of "*n'es-ce pas?*" We've all heard Americans say "*n'est-ce pas?*" in the course of English-language conversations, I'm sure. There's nothing wrong with it, but it sounds a bit affected. I'm for *enty?*

Mr. Stevens's reference to Gullah reminded me of a paper I'd written in college many years ago on Gullah. One of the sources I used was a small volume whose name and author I can't remember, but I'd love to get a copy of it. I recall an example of a conversation in Gullah that appeared in the book:

"Shum?"

"Shum!"

This conversation can have a number of different meanings, depending on tense and gender. "Shum?" can mean, Did you, Do you, or Will you see him, her, or it?

J. L. Dillard, in his excellent book *Black English*, says, "This 'lack' of pronoun sex reference [which isn't a lack at all; a pronoun system doesn't "need" sex reference any more than a verb "needs" tense] is characteristic of Gullah and is widely found in the Caribbean's Afro-American dialects."

So *enty?* (obviously "ain't he?" I should think) can mean "isn't it?" "isn't she?" "isn't he?" and other variations featuring "hasn't," "doesn't," past tense, future tense, and so on. Extrapolating from the lack of indication of tense in *shum* (Dillard considers this sort of lack a feature of language, not the absence of a feature), we can assume that "That was a beautiful shot, enty?" and "She's a beautiful girl, enty?" would be equally legitimate.

Realistically, I'm not so naïve as to think a nice, sensible word like *enty* has much of a chance in today's world. The trend, by and large, seems to be toward polysyllabic obfuscation rather than simplicity. The other day I heard a police officer on television say, "The suspect was discharging his weapon at me, and I exited the premises."

At a time (a point in time?) when people talk like that, *enty* will probably get lost in the burgeoning verbiage.

It's hard to imagine what causes a man to say, "The suspect was discharging his weapon at me." In the first place, in the mind of what I, for one, might consider a normal person, the "suspect," on discharging his weapon just once, would be transformed instantaneously from "suspect" to at least "S.O.B." And my normal person wouldn't "exit the premises" so much as get the hell out, enty?

I'm going to try using *enty* in casual conversation. I'll bet people who've never heard the word before will know exactly what it means.

That's a good kind of word.

November 6, 1973

SUBTLETIES OF THE TRADE

After reading a recent *Light Refractions* column that dealt with some peculiarities of descriptive sizes in salesmanship (e.g., Super Colossal olives and "Full Half Quarts" for pints of beer), a friend called to suggest I say something about women's clothes, speaking of weird size designations. I said I didn't know what he meant, and he told me that a size-twelve dress used to be one of the smallest, but to make women feel more petite the garment industry started calling size twelve size ten, and now the smallest sizes are six and even four. It's not that the real sizes of the dresses have changed, but the industry has fiddled with the numbers so that a woman who once wore a size fourteen might, without losing weight, now wear a size twelve.

I checked this story out. The people I talked to who are in the women's clothing business said it isn't so. But the evidence says it *is* so. Practically all I know about the dress business is what I've seen in *Save the Tiger* and a few other flicks: so I'm not grounded in the subtleties of materials and styles. My conclusions might be false, but I know that

I've talked to several women who've gone from size twelve to size ten while gaining a pound or two here and there. For years I've heard women refer to one another as "a size twelve" or "a size sixteen," and it's meant next to nothing to me. Now I'm beginning to get the hang of it. I know, for example, that uneven sizes—threes, fives, sevens, nines, etc.—are junior-miss sizes. But since there are a great many grandmothers wearing junior-miss clothes, I am more and more convinced that the size labeling of ladies' garments is, at least in part, a con job.

One of the offices I visited had an old-fashioned dress form with a partially formed dress on it. The dressmaker told me the dress was a size six. On the neck of the form was clearly stenciled Size 8. "I see," I said. "You're making a size eight and telling your customer she's a size six, no?" "No," I was told, "I'm making the waist very snug; so it's really a size six." I questioned that logic and got an explanation involving new fabrics that stretch and breathe and whatnot. I wasn't convinced of anything except the woman's absolute sincerity. She flatly denied that sizes had been redesignated to flatter the vain, but in the course of our discussion she mentioned that a size-eight model used to have a twenty-five-inch waist but today has a twenty-six-and-a-half-inch waist. She dropped that nugget while explaining that women are slimmer than ever, and that's why size twelve isn't as small as it once was.

Marva Shearer, the tennis player, is sure that dress sizes have been reassessed. She told me she'd been sent a selection of tennis dresses from a New York manufacturer. She'd told him she wore a size eight or ten. She chose two of the dresses, both of which fitted perfectly and both labeled Size 6. I don't believe that was a mistake. I think it was deliberate. I think that somebody figured out some time ago that if Mrs. America bought size-fourteen clothes at the Bon Ton but was a size twelve at La Femme Soignée, she'd feel happier at La Femme Soignée. Mrs. America would shift her custom to La Femme Soignée. It didn't take long for the Bon Ton to get the point.

As I said, I'm not grounded in the subtleties of the trade, but the evidence is pretty clear.

I think there's a comparable phenomenon in men's clothing. They can't mess around much with ordinary shirt and trouser sizes. They're measured in inches—waist, inseam, collar, and sleeve length. But athletic shirts—sweat shirts, tennis shirts, and so on—are measured small, medium, large, and extra large. No one with any sense of proportion could call me large. In fact, until a few years ago, I was a perfect example of "medium height and build." Today, with a younger generation that—in spite of white sugar and unnatural additives in its diet—has grown like giant sunflowers, I feel like a shrimp. Still, medium sweat shirts are small on me. For years I've felt that the manufacturers of sweat shirts want the average man to feel that, from the waist up, he's Herculean. The question is, if a large man needs an extra large sweat shirt, where does a *really* big guy—say, a tackle on the Dallas Cowboys—get his sweat shirts? And are they as Super Colossal as an olive?

December 4, 1973

-ISH

There's a radio commercial advertising a patent remedy for stomach distress—what the commercial calls "that fullish feeling." When I first heard it, I thought immediately of several words I'd find less fulsome than *fullish: full, stuffed, overfed.* I stopped just short of *bloated.*

Until then, the only research I'd done on *-ish* was for a Double-Crostic in which I wanted to use Ish Kabibble, who used to do fullish things for Kay Kyser's Kollege of Musical Knowledge and whose real name, I learned, is Merwyn A. Bogue.

"That fullish feeling," as a phrase, put me into a fit of mild depression.

I remember feeling that same depression years ago when someone asked me to arrive "around seven-thirtyish." But I'd never taken the trouble to check on what the top authorities said about *-ish.*

Webster's Third New International Dictionary says:

(1) of or belonging to—chiefly in adjectives indicating nationality or ethnic group (Finn*ish*) (Gaul*ish*) (Turk*ish*) (2a) character-istic or typical of (boy*ish*) (London*ish*), having the undesirable qualities of (amateur*ish*) (mul*ish*) (b[1]) having a touch or trace

of (summer*ish*), somewhat (purpl*ish*) (lat*ish*) (b[2]) having the approximate age of (forty*ish*) (b[3]) being or occurring at the approximate time of—esp. in words formed from numerals indicating an hour of the day or night (five*ish*) (eight*ish*).

So WNI III has accepted *fiveish*. I suppose they had to. Too many people use the form for it to be ignored.

As I noted, I was vaguely depressed when I first heard someone say "around seven-thirtyish." Then, like many people, I guess, I started saying *fiveish* and *eightish* as a joke—a quasi-snob form of humor, as one might say, "I was jes' funnin' ya," and say it often enough so that in time he begins to wonder to what extent he *is* funnin' himself. So a few years ago, when I caught myself saying things like *fiveish* without an inner chuckle, quasi-snoblike, I expunged that use of *-ish* from my vocabulary. I'll never use it again, no matter what WNI III says about it.

For some reason, my father hated the use of *-ish* except as an indication of nationality, and he didn't particularly like it even in that usage. He'd rather call a man a Briton or a Finn than say he was British or Finnish. I think he felt that *-ish* on any adjective had a tentative ring to it, as though to call one Finnish meant he was a Finn to a certain extent. *English* as a noun, of course, was okay. But he wouldn't even say "foolish." The adjective was *damfool*, as in, "That's a damfool thing to say!!"

I remember, when I was a boy, describing a dog as "sort of brownish." My father, a large and often florid man who was not one to remain calm in the face of such a damfool phrase, roared that "sort of brownish" didn't mean a damned thing, especially in describing a dog. He said I could at least say "light brown" or "dark brown." "Well," I said, "he was sort of grayish brown." "Ish!" he said. "For

God's sake!" He longed, understandably but unrealistically, for a simplified, clearly delineated, *-ish*less world.

WNI III's definition for *-ish*, "having the undesirable qualities of," came as something of a surprise to me. I'd never thought of *-ish* as being pejorative, but of course it usually is. *The Random House Dictionary* gets more specific:

> One of the common meanings of *-ish* is derogatory; that is, it indicates that something has the bad qualities of something else, or that it has qualities similar which are not suitable to it: *childish; mannish* (of a woman). The suffix *-like*, in the formation of adjectives, is usually complimentary: *childlike innocence; godlike serenity.* In an adverbial function it may be slightly disparaging: *Manlike, he wanted to run the show.* The suffix *-ly*, when it means having the nature or character of, is distinctly complimentary: *kingly; manly; motherly.*

The only exceptions I can think of to the rule that *-ish* denotes an undesirable quality are *boyish* and *girlish*, both of which are almost always followed by *charm*; and E. Y. Harburg did a whole number based on *-ish* in *Finian's Rainbow:* "Something sort of grandish . . . so sugar candish . . . Skies could be so bluish blue. Life could be so love in bloomish, if my ishes could come true."*

Nothing pejorative or put-down about *that* use of *-ish*, but I have a feeling that when Mr. Harburg wrote "something sort of grandish" he was getting *his* inward chuckle at the then-burgeoning use of *-ish* in the *fiveish* and *eightish* sense. If you're familiar with his lyrics, you know that he handles words the way the Globetrotters handle basketballs. In "When I'm Not Near the Girl I Love, I Love the

---

* "Something Sort of Grandish" © 1947. Chappell & Co., Inc.

Girl I'm Near,"** he's got two lines that I'll always relish (*-ish!*): "When I can't fondle the hand that I'm fond of, I fondle the hand at hand" and, "When I'm not facing the face that I fancy, I fancy the face I face."

I think that if E. Y. Harburg wanted to, he could sell me on "that fullish feeling," but even he couldn't have sold my father on it.

January 12, 1974

** "When I'm Not Near the Girl I Love I Love the Girl I'm Near,"
© 1946. Chappell & Co., Inc. Copyright renewed.
Both reprinted by permission.

YAY, TEAM?

During the hearings concerning the confirmation of Rep. Gerald Ford as Vice President, Ford commented, "You don't go out and tackle your quarterback once he has called the play," to which Sen. Harrison A. Williams, Jr., countered, "If your quarterback was running toward the wrong goal line, wouldn't you tackle him?"

"Yes," said Ford, "but that would be the exception rather than the rule."

The football metaphor has become an accepted commonplace in recent years. In my opinion, it's a bad one. It's simply inept.

Literate people are sensitive to jumbled metaphors. "The arms of the American Minuteman will be the scourge that stems the rising tide of vermin swooping down on the sleeping giant of outraged citizenry" collapses of its own disharmony, but "You don't go out and tackle your quarterback" may sound okay but it doesn't do the job.

The trouble is that football, though it's big business, is still basically a game. It is not comparable to government. To use football as a metaphor for government—and particularly for war—is to oversimplify, perhaps with deadly results. It's easy to see why football terminology is seized upon by politicians. It has a simple, pragmatic, virile ring to it. And I suppose everyone who makes money from football, with the exception of some of the players, has in one way or another fostered the idea that football is contained warfare. "He hasn't used the bomb yet" means merely that the quarterback or the coach or whoever really calls the plays has not yet called for a long forward pass. And have you ever watched those films of the glorious moments from the preceding week's games?—slow-motion pictures of enormous bodies hurtling high in the air and landing on their heads to the accompaniment of tympani and Götter-dämmerung-oriented music. The narrator, with a deeply resonant voice, sounds like the same one who used to say, "But France's military might crumbled rapidly before the invincible onslaught of Hitler's Wehrmacht" over shots of thundering panzer divisions. During the Vietnam war, there were constant references to "our team" and "our quarterback," and I once heard a man urging a hawkish policy say, "When you're on your opponent's five-yard line, you don't punt!" He didn't mention that, in football, neither do you saturate your opponent's city with high explosives from five miles in the air. Killing is equated with a game in this metaphor, and now there are a lot of people who deal with the fortunes of the American political system in the same terms.

The hackneyed, timeworn old Ship of State works much better. There is a real matter of life and death in the fortunes of a ship. I'd feel more comfortable with a man who said, "You don't get rid of your captain once he's set his course" than with one who used quarterbacks calling plays. If it's discovered that the captain's chosen course leads to Suicide Shoals, or that he and some of his fellow officers have been hacking holes in the hull below the waterline, it will occasion a greater sense of urgency than if the quarterback chooses to try for a field goal instead of a first down.

I hope the new Vice President will get over that "quarterback" business.

I think we have a tendency to think of the world in terms of winning, losing, happy endings, unhappy endings, and that sort of thing, as though the world were a game or a stage and all the men and women merely players. I'm a frequent listener to listener-response radio, as well as an ardent Letters-to-the-Editor fan. The other day I heard a woman who called in to one of the local talk stations say she'd like to live to be at least 100 "because I want to see how it all turns out."

World without end is a tough conception, but it doesn't help to picture the world in terms of opening kickoffs and final guns. In fact, thinking in terms of final guns, we just might *get* final guns.

January 26, 1974

SUPPLE STUFF

I was inundated by mail in response to the column suggesting "enty?" as an American equivalent to the French "*n'est-ce pas?*" The letters are still coming in, though the flood has slowed to a trickle. (Incidentally, several letters said things like, "Perhaps you or one of your staff could. . . ." This might be as good a time as any to leak the fact that it's only me. No staff. That explains in part why too many letters that have me saying, "I've *got* to answer *that* one!" never get answered. It's always me against the deadlines, and the deadlines win.)

A surprising number of people said that they and/or family and friends have used "enty?" for years.

Many readers pointed out other regional equivalents for "*n'est-ce pas?*" or "enty?" I learned from several people that in Wisconsin "and so?" "ainna?" "enna?" "entso?" and "inso?" are common. In the Pennsylvania Dutch communities, a simple "ain't?" serves.

There are supporters and opponents of the Canadian "eh?" I feel uneasy with "eh?" It conjures up a codger with

an ear trumpet saying, "Eh? What's that, young feller?"
No doubt a throwback to the funnies of my youth.

By far the most frequently touted equivalent of "enty?"
is "right?" Dozens of people questioned the need for
"enty?" or anything else while the ubiquitous "right?" is
already on the job.

"Right?" doesn't quite make it, I think. My own feeling
is that it's graceless. I'm now aware that a great many people find "enty?" even more graceless, but "right?" to me is
like a verbal elbow dug into the ribs.

There's also "no?" as an answer to the need for a *"n'est-ce
pas?"* but "no?" for some reason speaks to me with a south-
of-the-border accent: "I theenk your Chiquita has eyes for
the Greengo, no?" Another throwback, I suppose, this time
to the low-budget westerns of my youth.

My favorite is still "enty?" with "enna?" moving up fast
on the inside.

I particularly liked the letter from Jake Lee, the noted
aquarellist. Mr. Lee, like Dong Kingman, is American-
born of Chinese parents. He starts his letter outrageously:
"The opening words when a group gathers around a poker
table are, 'Everybody enty?' "

Lee studied painting under Kingman. He continues,
"Years ago when Dong Kingman lived in San Francisco,
he used to stand around the corner of California and Mont-
gomery Streets at 4:30 in the afternoon and watch the
girls come out of the office buildings. As the wind picked
up the skirts . . . Kingman would mumur, *'Fow goy.'*
You see, Kingman can't whistle so he couldn't resort to
the usual wolf call. *Fow* [rhymes with *wow*] *goy* literally
means 'supple stuff.' I think the wolf whistle is a great
American expression and should be incorporated into our

language, but there is no spelling for a whistle. Since we are expanding our cultural exchange with China, why not trade a wolf whistle for *fow goy,* enty?"

Jake Lee is right: There is no spelling for a whistle. And the wolf whistle is a uniquely American expression, as far as I know. It's a remarkable phenomenon, really. It has a basic structure within which innumerable interpretations are possible, from a terse couple of blasts signifying simply "Hi, Babe!" to a sensuous paean of appreciation with insinuations of sultry evenings in Acapulco.

It's hard to imagine an American who doesn't know what a wolf whistle sounds like. It has its accepted form, and we all recognize it, whether it's performed by a masher or a myna.

I think, though, that the idea of a cultural exchange—wolf whistle for *fow goy*—is unworkable. For one thing, anyone standing on a Peking street corner whistling at earnest young socialist women wearing pajamas and absorbing the thoughts of Chairman Mao would almost certainly be deemed soft in the egg roll and banished to the boondocks and a life of digging irrigation ditches.

As for *fow goy* in the United States, I'm sure that would go over only in a place like San Francisco, where eccentrics are valued, and even there only if your parents were Chinese.

If you can't whistle, I think you'd do better to say something like "Wow!" (rhymes with *fow*) and, especially now that the joys of Yiddish are available to everyone, leave the *goy* out of it entirely, enty?

February 9, 1974

HOPEFULLY

These days, it seems to me that *hopefully* is being inveighed against almost as much as it's being used. Many linguistic conservatives consider "Hopefully, the average American's cabin cruiser will not be a victim of the energy crunch" to be an abomination, since the average American's cabin cruiser cannot be hopeful and certainly can neither be nor not be a victim hopefully.

Well, a crime against the English language is not to be taken lightly, so I've given *hopefully* some serious thought. Time that I might have spent trying to solve the Palestinian refugee problem or to develop an automobile engine that runs on salt water has been used in pondering *hopefully* and the degree of evil thereof.

Webster, both second and third editions, gives little or no justification for *hopefully* in the sense of "it is hoped that" or "I hope"; nor does the Oxford English Dictionary. Random House does though. Its second definition reads: "It is hoped; if all goes well: *Hopefully, we will get to the show on time.*"

At the risk of alienating some friends, I think I have to

come down on the side of Random House for one over-whelming reason: Too many people use *hopefully* to mean "it is hoped; if all goes well." I don't mean people who use sloppy English regularly, but well-educated, erudite peo-ple who obviously care about the language—some of our most prestigious news commentators, for instance. And at least one author-cum-English professor, John Gardner, who, in reviewing Kenneth Koch's book in *The New York Times Book Review*, wrote, "*Wishes, Lies, and Dreams* tells how he did it and how, hopefully, anyone can do it."

I suppose he could have meant that anyone who's im-bued with hope can do it, but I doubt it.

"The Congress, hopefully, will pass legislation limiting the President's power to tax indigent widows" may be ambiguous. Does it mean, "It is hoped that the Congress will pass . . ."? or does it mean, "The Congress, hopeful that the President won't veto or something, will pass . . ."? I've thought of a few other examples of possible am-biguity, but they all boil down to the same structure—an unlikely one for any meaning other than the Random House "it is hoped."

I'll admit that the formal inconsistency in that relatively new meaning bothers me. *Hopefully* should mean "in a hopeful manner." I admit, too, that I don't think I've ever used *hopefully* in the sense of "I hope" and that when I hear or see it in that sense, a little click ("there it is again") snaps in my head, as, for instance, my noticing it in Pro-fessor Gardner's book review. If I'd been really comfort-able with the new *hopefully*, it wouldn't have registered.

The point is that purists can't win in a case like this.

A couple of years ago, one of my favorite thinking news-men, Robert Abernethy of NBC, said, "We say, 'Hope-

fully, I'll see you tomorrow.' What that means is when I
see you tomorrow I will be hopeful. But who can guarantee
that? What we mean to say is: I hope I'll see you tomorrow.

"I don't mean to be petty, but the language is worth defending. Sloppy speech usually implies sloppy thinking,
which can be dangerous."
I cheered him at the time and I cheer him now, but while
there may be some point in fighting against overwhelming
odds, trying to whistle a typhoon to a standstill is just a
waste of time and energy.

As long as there are enough of us who agree that "the
language is worth defending," we're not in such bad shape,
but there's no denying that things do change. Not many
years ago, *contact* as a verb was in bad repute. Today,
both Webster's Third and Random House have accepted
it without reservation. They had no choice. It was being
used as a verb, not just by too many people, but by too
many literate, influential people.

Fun, once either a noun or a verb, is now either a noun,
a verb, or an adjective. I bridled when I first heard,
"They're a fun couple," and I still do. But there's no reason
why *fun* shouldn't be an adjective. To *have* fun requires
that *fun* be a noun, but there's nothing wrong with "Skiing
is fun," where *fun* seems to me to be much more comfort-
able as an adjective.

Maybe someday I'll talk about a fun couple I know, and
it's more than likely that someday I'll use *hopefully* to
mean "it is hoped." I hope Robert Abernethy, Laura Z.
Hobson, et al., will forgive me. I say that most hopefully.

February 23, 1974

WATCH YOUR LANGUAGE

" **A**s more facts came to light indicating that Vice President Agnew had been uncommonly prestigious as governor of Maryland, it became increasingly clear that he would have to resign."

I maintain that that sentence is absurd, and I said so a few weeks ago on a television interview.

It makes sense if you're using *prestigious* as it's defined in *The Oxford English Dictionary*: "practicing juggling or legerdemain; of the nature of or characterized by juggling or magic; cheating, deluding, deceitful; deceptive, illusory."

I was trying to make a point I've touched on in previous columns: that language changes, and that the most intelligent way to handle its changes is to accept them while being conscious of them. But, almost predictably, a woman called in to NBC after the interview to say that *prestigious* still means what the OED says it means and that I didn't know what I was talking about.

Webster's New International Dictionary, Second Edition, gives "conjuring" and "pertaining to, or of the nature of, conjuring or magic" as obsolete meanings for *prestigious* but is in general agreement with Webster's Third and Random House on the current meaning: "honored, esteemed."

The woman who called NBC reminded me of someone who wrote me several years ago to say that she liked Double-Crostics but that she didn't like mine because I didn't use the English of Shakespeare and Wordsworth. I think I sent an answer to her letter. (I'm not sure, because I answer lots of letters in my head, committing them neither to paper nor to mailboxes. These mental replies have more reality to me than to my correspondents.) What I said, in essence, was that Wordsworth didn't use the English of Shakespeare, nor did Shakespeare use the English of Chaucer, and that that was just as well, because if he had, Dick Burbage, being a speaker of Shakespearean English, would have had a lot of trouble with the lines.

The Oxford English Dictionary is a remarkable work, but, lexicographically speaking, it's old. My edition is dated 1971, but it's a compact version of the pre–World War II twelve-volume edition.

The next edition will have to include the new meaning of *prestigious*. The current edition says "now *rare*." It says of *prestige* that it's from the Latin *præstigiæ* ("illusions, juggler's tricks"). The second meaning it gives for *prestige* is "blinding or dazzling influence; 'magic,' glamour; influence or reputation derived from previous character, achievements, or associations, or esp. from past success."

I think the woman who called NBC should rejoice with me. Here old *prestigious*, a word that was said to be rare about forty-five years ago, was languishing on the shelf and is now making a whopping comeback, rooted firmly and happily in the relatively new meaning of *prestige*. This is a heartening example of a has-been word, with nothing to look forward to but the ignominy of sinking from *rare* to *obs.* or *archaic,* resurrected to new vitality.

This demonstrates as clearly as anything else the vitality of language and the two-way function of dictionaries. The earlier dictionaries were compiled in an attempt to establish fairly immutable standards in language. After a time, it became obvious that language could not be held at a standstill. As new generations change the look of the world, so do they change their ways of speech. Dictionaries serve as authorities at the same time as they serve as reporters of what, despite their authority, is happening to the language.

That's one of the features that make *The Oxford English Dictionary* so great. Not only do its lexicographers realize that the meanings of words change: They set down a most informative record of each word's history, with examples of its use by various authors at different times. So the giant old OED may not be up-to-date (obviously, it contains none of the vast numbers of technical words coined since it was compiled, such as *iconoscope, transistor,* etc.), but it deals in depth with those words it includes.

Among other things, it proves the point that it's all right to be a linguistic conservative, but it's possible to carry conservatism to such an extent that your contemporaries won't know what you're talking about.

March 9, 1974

MENTAL TEMPLATES

Linnie Whorf moonlights by typing my puzzles into respectable shape for submission to SR/WORLD, *The New York Times*, and Simon & Schuster, which publishes the puzzle books. Linnie is a puzzlehead. In fact, she comes from a family of puzzleheads. She tells me that, during her formative years, anyone in the company of her three brothers, her mother, and, above all, her grandmother had better not be inept at word games, because that was what life was all about. Linnie's regular job is with a firm bearing the impressive name Environmental Dynamics.

She phoned me a couple of weeks ago to say that she had some puzzles typed up and ready to go. She also said that the firm's accountant had been in the office that day, had discovered Linnie's puzzleheadedness, and had admitted to being a puzzlehead himself. "He gave me a list of numbers," Linnie said, "and asked me to figure out why they were in that sequence." She said she couldn't figure it out. I told her I didn't have time to fool around with number puzzles these days, but just out of curiosity I asked her what the sequence was. "Eight, five, four, nine, one, seven, six, ten, three, two," she said. "I told him right away that the first two pairs of digits each added up to thirteen, but he said that had nothing to do with it. It's driving me nuts."

"Well, I don't have time for that sort of thing," I said. But I wrote the numbers down, and they started driving *me* nuts.

I tried adding, subtracting, multiplying, and dividing. Soon I figured that it wasn't a math puzzle but, somehow, a word puzzle.

I wrote down 8, 5, 4, 9, 1, 7, 6, 10, 3, 2; eight, five, four, nine, one, seven, six, ten, three, two; VIII, V, IV, IX, I, VII, VI, X, III, II.

If you're a fellow puzzlehead, you'll understand the sort of absurd reasoning processes I went through. If you're not, perhaps I can introduce you to a weird insight into this peculiar insanity:

"Eight five four nine."
"Ate five for nine?"
"Ate five for nine one seven six."
"Ate five for nine i seven six?"
"Ate five, for nine is even six?"
"At five, for nine is evens? ix ten? 9, 10?"
"Is even" sounds good. I'm on the right track.
"VIII V IV IX is even."
"VII IV IV IX is even?"
"7, 4, 4, 9 is even."
"26 is even six ten three two."

At this point, I begin to lose my mind. "Let's see: The final two digits, in Roman numerals, are III, II. I, i, I, i, I! Maybe. Why not? It sounds like a line from 'The Goldbergs' or a variation on '*Ay caramba!*' or '*Ay, Chihuahua!*'

Shades of Leo Carrillo: '*Si, si, Ceesco.*' I'm on the wrong track, obviously."

Back to work. But always there's a nagging presence lying there on top of a stack of magazines: that damned scrap of paper and its nagging 8, 5, 4, 9, etc.

At 3:00 A.M. a dog barks. I wake up, and instead of going back to sleep, I go back to: "Twenty-six is even. Six ten three two. 26 is evens ix, 10, 3, 2. 26 is evens 9, 10, 3, 2. 26 is evens. 24!" I'm losing sleep, falling behind in my work, and going bananas.

For about a week and a half, that sequence elbowed its way into my mind, throwing everything else off stride. A few days ago, Linnie came by with a few freshly typed puzzles. I asked her if her accountant friend had given her the answer to the number sequence. "Yes," she said. "They're in alphabetical order. Isn't that infuriating?"

"That's ridiculous!" I said. Then my mind started snap-crackle-popping like high-voltage Rice Krispies, and I realized that, while it was indeed infuriating, it was not at all ridiculous. It was a marvelous example of the way our minds (or my mind at least) can block out the obvious because we have built-in ways of seeing things.

If, instead of eight, five, four, nine, etc., it had been eggs, fettuccine, flour, nutmeg, onions, soap, steak, tea, tomatoes, turnips, I think I'd have seen the alphabetical order almost immediately. But, because numbers have their own inherent order, it's difficult to shift gears and see them, not as numbers, but as words with alphabetical properties. After spelling the numbers out, I should have caught on.

A few years ago, a corporation—I can't remember which one—had full-page ads featuring the sort of problems that necessitated this temporary suspension of the mental templates—a shelving of the old gestalt. They were fascinating lessons in problem solving.

I wish I'd borne those lessons in mind a couple of weeks ago. I feel pretty silly having racked my brains all this time over 8, 5, 4, 9, 1, 7, 6, 10, 3, 2 when all the time it was literally as easy as ABC.

March 23, 1974

ON KNOWING
WHAT'S WAT

Good show, what?"
"What, no marshmallows?"
"What do you take me for, a fool?"
"What are you taking me for, a walk?"
"What are you taking me for, ransom?"
"What the hey?"
"What a word!"

My favorite experience with *what* happened several years ago when Andy Van Scoyk was three years old. Andy's one of the sons of Robert Van Scoyk, the television playwright. I was staying with the Van Scoyks, and Andy's father and I were having breakfast when Andy came into the room, knelt on the chair next to mine, and said, "Daddy, you know what?" "No," said Bob. Andy gave me a smirk of triumph, as though he'd just won another round, and said, "Daddy never knows what!" Then he asked, "Do *you* know what, Tom?" Just for the hell of it, I said, "Sure I do, Andy."

If I'd flapped my ears and flown around like Dumbo, he couldn't have been more astonished. After a couple of seconds, he scrambled down and ran from the room, rapturously shouting, "Mommy! Mommy! Tom knows what!"

There's only one reasonable response to "Do you know what?" It can be phrased as "No" or as "What?" or as "No, what?" but the answer "Yes" carries with it implications of omniscience far more impressive than knowing beans or one's onions. *What*, in fact, carries the germ of limitless knowledge.

I once had a *what* experience that would have had Abbott and Costello drooling.

It was in an apartment in New York. Among the guests was a son of Emperor Haile Selassie. His date, who had cooked the dinner, was a lovely young Ethiopian woman, and the dinner was a delicious sort of stew whose ingredients I've long since forgotten. I remember that it was accompanied by a red sauce, the makings of which she'd had sent to her from Ethiopia. I think that in Ethiopia those ingredients were used mostly for cauterizing wounds and branding cattle, but, as a sublimated masochist, I love peppery foods and I won the admiration of the prince and his date by asking for more. The only things I remember about the stew are that it contained some chicken and that it was called *wat*.

I'm sure you could write the conversation yourself, but here's an authentic version attested to by me:

Self: This is delicious.
Ethiopian Girl: Thank you.
Self: What do you call it?
E.G. Wat.
Self: This dish.
E. G. Wat.

Self: The stew dish. What's its name?
E.G. Yes
Self: No. I mean what is the name of the stew?
E.G. That's right.
And so on.

She played me skillfully and well, to the great delight of our hosts and the young prince of Ethiopia.

Abbott and Costello's great infield of Who, What, I-Don't-Know, and I-Don't-Give-a-Darn (I suppose it was I-Don't-Give-a-Damn on the stage, but I saw the film version) never displayed more razzle-dazzle.

It may be significant that What, being able to handle the various chores of neuter interrogative pronoun (singular and plural), adverb, adjective, noun, interjection, and being the repository of virtually infinite knowledge, was at second base, otherwise known as the keystone sack.

What else?

April 20, 1974

HOPEFULLY REVISITED

Sometimes I know that a column is going to elicit a strong reaction. The one a couple of months ago on *hopefully* was one. What I didn't expect was such a *mixed* reaction.

You may recall that it dealt with the use of *hopefully* meaning "it is hoped that" or "I hope that," rather than "in a hopeful manner." To my surprise, I got letters thanking me for taking a strong stand against this usage, letters excoriating me for accepting the usage, and letters asking what the fuss was about.

It's certain that the overwhelming majority of English-speaking people have not suspected that there was such a thing as a *hopefully* problem, and I have a feeling that a substantial majority of even such a discriminating group as *Saturday Review/World* subscribers have accepted *hopefully* as "it is hoped that" without a qualm.

Two readers, Henry A. Weber, of Paradise, California, and Jaroslav Pelikan, dean of the Yale Graduate School, suggested that *hopefully* is an invasion into English of the German *hoffentlich.* That's possible, I suppose, but I don't think it's probable. I think it's just sloppy English. I think it's an elliptical way of saying "I suggest hopefully that," or some such thing.

The *Random House Dictionary* accepts this usage. So does *Webster's New World Dictionary*, Second College Edition, which adds, "regarded by some as loose usage, but widely current."

I said before that the dictionaries have to accept this usage for the obvious reason that it *is* "widely current."

Many readers roasted me for not keeping the faith. Mr. Paul Van Gieson, of Hillsdale, New Jersey, expressed the very common feeling that I "fizzled away to the can't-fight-city-hall attitude."

The point, I think, is that it's not a question of fighting city hall. Language is not a product of legislative fiat. If the city council had *voted* to make *hopefully* mean "it is hoped that," I'd be in the van of the drive to toss the bums out. But that's not how it happens with language.

"Hopefully, he'll get a job."

From a logical point of view, that means that he'll get a job—no doubt about it—and in a hopeful manner (hopeful of being a success in the job, perhaps). But we know that that's *not* what "Hopefully, he'll get a job" means. It means that he may or may not get a job, but it's hoped that he will. If you could pass a dozen laws against that meaning, it wouldn't change a thing.

Mr. Van Gieson goes on to equate *hopefully* with "all the Haldeman-Ehrlichman-Nixonmanisms like 'at this point in time,' 'I want to emphatically say,' 'we do not want to self-destruct,' etc." I'm glad he mentioned that, because I think there's a fundamental, important difference here. *Hopefully* for "it is hoped" is a form of ellipsis—a linguistic short-cut. Those "Haldeman-Ehrlichman-Nixonmanisms" are pompous gobbledygook. They're designed to give a highfalutin aura to trash and to make nonsense sound legitimate. I have a very strong feeling that the overwhelming majority who never gave a thought to *hopefully* reacted to the linguistic absurdities of the Watergate crowd with distaste, bemusement, laughter, or a combination of the three.

"Hopefully, the price of beef will come down in the coming months" has nothing to recommend it logically, but linguistically, it's a *fait accompli,* whether we like it or not.

Language is full of illogicalities. One that comes to mind immediately is *proud.* It's never made sense to me that a man can be proud of his art collection and that his art collection can be his proudest possession. How can an art collection be proud? How can it be prouder than a brand-new stereo system? The art collection and the stereo system are proud to the degree that their owner is proud of them, respectively. That's not logical, but it's an inescapable fact of the language.

There are many misuses of the language I'll fight against —most of them because they are dishonest or somehow misleading. But, though I prefer *hopefully* as *hopelessly's* antonym, and I think I'll continue to use it that way, I refuse to tear the epaulets from anyone's shoulders for saying, "Hopefully, the price of beef . . ."

May 4, 1974

MINDLESS CLUTTER

Charlotte F. Moss, of New Haven, Connecticut, is one of several people who have written to me about the inane expression "I could care less." As everyone knows, it is used in virtually every instance to mean its very opposite.

On the basis of no hard evidence whatever, I believe "I could care less" is a British import. I have a vague and possibly erroneous memory of having first heard it used by some English tourists in Paris in about 1953. If I'm right, it's been around for a couple of decades, at least.

I don't think it's worth worrying about.

"I could care less," "You better believe it," "That's for damn sure," and so forth are mindless clichés that clutter up today's language. Yesterday's language had its own junk: "You said a mouthful," "Don't take any wooden nickels," "Ain't it the truth?" They don't mean any more than "Hiya" or "Yeah" or "Good luck." "I could care less" and "I couldn't care less" both mean approximately "So what?" which in turn doesn't mean much more than a shrug or a grunt.

Shrugs and grunts will survive, and I suppose they'll always have their verbal counterparts, but I'm sure it won't be long before "I could care less" and "You better believe it" are as dead as "Twenty-three skiddoo," "Hotchacha," and "Yea, forsooth."

Ms. Moss, having mentioned that "I could care less" means its opposite, went on to remind me of the song from a few decades ago about the spurs that go jingle-jangle-jingle. "They sing, 'Oh, ain't you glad you're single,' and that song ain't so very far from wrong."

That, too, means its opposite.

"What's clearly meant," says Ms. Moss, being more grammatical than the lyricist, "is 'the song isn't very far from right.' "

Most people probably don't pay much attention to the lyrics of popular songs, but if you've tried writing lyrics, you know how tough it is. You respect those who write good ones, and chances are you listen more carefully than those who haven't tried it.

Those jingling spurs reminded me of another, much older, cowboy song—the classic "Bury Me Not on the Lone Prairie"—which I learned as a very young cowpoke in the badlands of southern Westchester County. Unless I learned it all wrong, the opening line is "Oh, bury me not on the lone prairie." Then, with a total disregard for dramatic unities, without even applying for a poetic license, the last line says, "When I die, you can bury me 'neath the western sky on the lone prairie."

There's no indication that the singer has changed his mind. It's just that by the time he gets to the punch line, he's forgotten his opening statement, and he blithely expresses its contradiction.

You can get away with almost anything if you set it to music, which is undoubtedly one of the reasons singing commercials are so ubiquitous. You remember them, although they don't have to say anything beyond the name of the product.

As a child I'd sung, "It rained all night the day I left, the weather it was dry. The sun so hot I froze to death, Susannah don't you cry" for years before it struck me that it was absurd—primitive humor. I think I was probably unusually slow in the case of "Oh Susannah," but I bet we've all sung one song or another without thinking of the meaning.

If you're old enough to remember Pearl Harbor, you're old enough to remember "South of the Border." There's a line in that song that I still cherish. I'm not sure anyone else noticed it, but I did, and I'll share it with you: "There for a tender while I kissed the smile upon her face."

I maintained in 1943, and I maintain today, that *that* is just about impossible. If you've never tried it, give it a whirl. Ready? All right, Ladies, smile at the Gentlemen. Gentlemen, kiss the smiles.

If you're at all normal, you'll have noticed that the "tender while" turned into something quite different. Most likely uncontrollable fits of the giggles. You can't kiss a smile for a tender while. If one half of a couple is smiling and the other half is kissing, something's got to give. Kissing a smile upon a face results in virtually instantaneous loss of composure, dignity, and maturity. That's still better than worrying about caring or not caring less.

May 18, 1974

SMASHING

Mr. Elliott Graham, of New York City, writes:

I'm watching with horror the growing acceptance of that weary old British adjective *smashing*. It's been a long fight, but the fashion editors and the advertising copywriters have finally established what I feel is a permanent beachhead in this country. Now it turns up in interviews with all the new young film and television personalities who have made their first visit to London— and it ruins (for me) all kinds of unlikely news stories. I see no possible way to stunt its growth.

He may be right, but it's my feeling that *smashing*, like many other Anglicisms, travels badly. Not that it sounds worse in Pittsburgh than in Manchester, but that it sounds worse in an American accent than in an English accent. There are certain usages that are at home in English English and others that are at home in American English that don't do well when they go visiting.

In fact, I take back something I said about a month ago. I suggested the possibility of the adoption into American of the English interrogative "what?" as employed to such great effect by the late Nigel Bruce: "I say, Holmes—frightful, what?" It won't work in American, and the same goes for *smashing*.

My earliest memory of the use of *smashing* to mean "sensational" is of a postcard somewhere on the Côte d'Azur in the early Fifties. It had a drawing of a lot of bikinied nifties on a beach with the scrawled caption "The mamzelles here are smashing." Its meaning was clear from the context, but the word *smashing* struck me as bizarre at the time; so I bought the card and sent it to a friend in New York. Afterward, I heard *smashing* used constantly by Britishers in Europe—smashing girl, smashing day, smashing party. I liked it, but, except when imitating an Englishman, I don't think I ever said it.

It's a question of what fits, and I think it's largely discretionary. I had a cousin, born and raised in New York City, who in his early teens referred to "blighters." I don't know why. Perhaps he was too well-read for his age. He may have felt comfortable saying "nasty blighter" and "he's a decent blighter"; I never asked him about it, but it made me uneasy. It was as though George Wallace started complaining about someone's *chutzpah,* or Lieutenant Columbo said, "Listen, I don't wanna pry, but it looks like you had quite a *bagarre* here."

I once heard an Englishman described as "a very huntyfishy-shooty sort of chap." That's a phrase that can be used only by a certain sort of person—one not merely English, but a particular *type* of English. Or the description of a young Englishwoman: "very jolly-hockey-sticks." Now, 999,999 out of 1,000,000 English-speaking people would sound ridiculous saying that, but when I heard it, it sounded just right, because it was said by the right sort of person. Certain expressions should be limited to the people who know how to handle them.

A fellow named Amarillo Slim has had a lot of publicity in the last couple of years as a champion poker player, all-around gambler, and colorful character. When he was asked how it felt to drop $50,000 in one poker game, he drawled, "It shore put an early frost on mah peach trees." An excellent reply, but only for one with the proper style. I have no idea whether Amarillo Slim owns peach trees. It's not important. Terry-Thomas can talk about a smashing performance, but he'd be all wrong trying "an early frost on mah peach trees."

Mr. Graham may be right. *Smashing* may have established a permanent beachhead here, but I doubt it. He bridles at it, as I do (though in my case it brings back memories of the Côte d'Azur, so it's not all bad). I think there's an element of affectation in *smashing* that is sensed by even the TV and film personalities who say it; so I feel that its beachhead is tenuous at best. We went through a period when everyone who had spent a couple of afternoons on the Via Veneto and then got on a TV talk show was saying "*Ciao*" to everyone.

Jet flight has mingled cultures and languages as never before, and you may call me provincial, but I think that the Americans who *ciao*'ed one another on nationwide TV are the spiritual sons and daughters of Fran Dodsworth.

It's been a long time since I've heard *ciao* on TV. Maybe I'm just watching the wrong shows, but I hope the Americanization of *ciao* has withered of its own fatuousness.

It could be wishful thinking, but I predict the same fate for the Americanization of *smashing*.

June 15, 1974

ᏬᏗᏬᏗᏬᏗᏬᏗᏬᏗᏬᏗᏬᏗᏬᏗᏬᏗᏬᏗᏬᏗᏬ

LINGUISTIC INANITIES

When I was in high school, an English teacher told me that "somebody else's" was incorrect and that I should say "somebody's else." I wanted to speak the language properly, and she was the best authority I had, so for a few years I said "somebody's else" instead of "somebody else's": "somebody's else jacket," "somebody's else girl friend," and so on. I finally backslid, on the grounds that "somebody's else," no matter how often I said it, sounded nonsensical. I'd never seen an else; how could an else be somebody's?

My teacher was correct at the time, but she didn't finish the lesson. *Webster's New International*, Second Edition (1953), says:

> When used with the prenominal expressions *anybody, everybody, someone*, etc., in the possessive construction, *else* takes the inflectional form of the possessive case, except that when the noun is not adjacent, the accompanying pronoun correctly, though now less frequently, takes the inflection; as, somebody *else's* hat; we do not know whose *else*; the hat can be nobody's *else*.

Today, "now less frequently" has become "now hardly ever," and almost nobody says "nobody's else."

I'm convinced that millions of people have been taught half-truths or downright absurdities by English teachers.

Years ago I was visiting friends. At one point, I said, "Sure, Sam, I'd love a cup of coffee." As he poured the coffee, he told me that "a cup of coffee" was bad English. Sam must have been in his fifties at the time, and some screwball English teacher of his youth had told him that "a cup of coffee" would be a cup made of coffee or a coffee-colored cup, just as a cup of gold is not a cup full of gold, but a cup made of gold or colored gold. If he meant a cup *full* of coffee, he should *say* a cup full of coffee.

I told Sam that his English teacher was full of malarkey and threw in something about "A book full of verses underneath the bough, a jug full of wine, a loaf of bread, and thou beside me. . . ."

It could have been faulty memory on Sam's part, but I doubt it. The English language evidently invites not only weird interpretations of existing rules but also the occasional dreaming up of fanciful rules, and there are pseudo-purists who demand strict adherence to rules that exist principally in their own minds. I once said, "People that say . . ." on television. Sure enough, NBC got a letter from a gentleman who said that anyone who says, "People that . . ." doesn't know much about the English language. While admitting that I prefer "people who" to "people that," I'll stand firm in support of both "The Man That Broke the Bank at Monte Carlo" and "The Girl That Married Dear Old Dad," and if Irving Berlin had written "The Girl Whom I Marry," no one would have sung it.

About a year ago I got a letter from a lady who, in the kindest way imaginable, took me to task for writing, "I don't think I've ever heard . . ." or something like that.

"Please!" she wrote, "You *do* think!" She said she enjoyed *Light Refractions,* so I admire her taste and intelligence; but I'll bet that she'd once been the victim of an English teacher who told her that "I don't think so" is wrong and "I think not" is right. The fact is, of course, that "I don't think so" and "I think not" do not mean precisely the same thing, and *think* has many meanings, among them "suppose" and "believe."

From time to time, I hear from people who are terribly upset by "aren't I?" I maintain that "aren't I?" is perfectly okay and that any English teacher who says it's not is looking for trouble. I'd like to see *ain't* given legitimate status, but short of that, I'll take "aren't I?" over "amn't I?" which is plain ugly, and "am I not?" which is fine for oratory but is stilted in everyday speech.

Checking my opinion with Henry Fowler, I find myself in the best company imaginable. He says, in *Modern English Usage:*

> Though *I'm not* serves well enough in statements, there is no abbreviation but *a(i)n't I?* for *am I not?* or *am not I?*; for the *amn't I* of Scotland and Ireland is foreign to the Englishman. The shamefaced reluctance with which these full forms are often brought out betrays the speaker's sneaking fear that the colloquially respectable and indeed almost universal *aren't I* is "bad grammar" and that *ain't I* will convict him of low breeding.

Much of good English is a matter not just of the rules but of their common-sense application.

There are millions of people who have been alienated from this most beautiful language by English teachers who insist on such inanities as "cups full of coffee" and "am I not?"

It's a damned shame.

July 13, 1974

ᐃᐧᐃᐧᐃᐧᐃᐧᐃᐧᐃᐧᐃᐧᐃᐧᐃᐧᐃᐧᐃᐧᐃᐧᐃᐧᐃᐧᐃᐧ

LANGUAGE
AND SURVIVAL

In 1924 the Washington Senators won their first and only World Series, *The Saturday Review of Literature* was born, and so was Little Orphan Annie (Leapin' Lizards! That kid is 50 years old!?!?).

The last 50 years are traditionally represented on those linear scales of the History of Man starting with the Pleistocene epoch as a little dot at either the upper or right-hand end of the line, depending on whether it's horizontal or vertical. We're told in the accompanying text that the most recent half-century was a real blockbuster in terms of increased population, increased pollution, increase in the number of scientists per capita, or what have you. And every time I see one of those scales it scares hell out of me, because its point is usually well taken and because the changes that have been taking place on our planet during the last 50 swings around the sun are staggeringly different quantitatively and qualitatively from anything that has ever happened before.

Little Orphan Annie is the only thing I can think of that has remained the same.

The most obvious changes in the English language have been the new words, I suppose, but there have been other, more subtle, changes.

In 1924 radio was just barely out of the crib. The hottest program on the air was "The Happiness Boys," Billy Jones and Ernie Hare, whom I still remember hearing, though some years later than 1924.

The Jazz Singer hadn't been made yet, and television was a small gleam in the eyes of GE, Westinghouse, RCA, and AT&T. The iconoscope had been patented the year before, in 1923.

The increase in English vocabulary attributable to World War II, the atomic age, space exploration (which, we are reminded, gave us Teflon), computers, and technology in general—with its proliferation of new terms for new inventions, discoveries, and processes—hardly seems worth mentioning, not because it is not significant, but because it is so overwhelming and so obvious.

More subtle effects on the language have been brought on by the results of technology itself. I remember a friend who had grown up in the Tennessee Valley during the Thirties. "Man, Ah teyeh," he said, "when thet air TVA come in, yeh jes' rair on back an' flep a swetch, an' on come the laht!" Today, his counterparts can flip a switch and hear Alistair Cooke or Walter Cronkite.

Fifty years ago, they might have gone to the movies and seen Tom Mix, Mabel Normand, or Gloria Swanson. Whoever it was they saw, the speaking voices they heard were in their own heads, and the accents were Tennessee Valley accents. Today, it's hard to imagine anyone in the United States who has not been exposed to the voice of Sir Laurence Olivier, even if only speaking about the earth-shaking new Polaroid SX-70 camera, whereas in 1924 only a relative few would have been able to imagine comparable speech.

I suspect that, although it's widely lamented that the average college freshman today can hardly tell a subject from a predicate, the average American speaks better English, with a much larger vocabulary and far fewer serious grammatical errors, than ever before, thanks to radio, TV, sound films, and the great increase in mobility since World War II.

The number of college-educated has been increasing faster than the number of well-educated, apparently, but the relatively *un*educated are better educated than they were 50 years ago.

The simultaneous upward and downward movements toward egalitarianism might make excellence seem not superior but extraneous and suspect; still, I don't fear for the future of the language. Certainly, it is threatened by a few super-egalitarians who preach that language is "only" a means of communication, so if the language communicates, however sloppily, it serves its purpose and is therefore all right.

That's a sound theory if all you need are gestures, grunts, nods, and strokes among people with relatively common goals and understandings, but the future of the world depends upon an infinitely greater degree of sophistication than that, and that sophistication requires that people communicate within a well-defined set of rules. Communication is difficult under the best of circumstances. Without rules it becomes next to impossible.

I don't fear for the future of the language because I'm certain that enough of us are aware of that fact and that our numbers will increase. Some of the rules may change as new needs arise, but rules there must be, and if we're not using just about the best set of rules available, we're putting ourselves at a terrible disadvantage.

There's nothing elitist about the proper use of language. It's a matter of survival.

August 10, 1974

⚬⚬⚬⚬⚬⚬⚬⚬⚬⚬⚬⚬⚬⚬⚬⚬⚬⚬⚬⚬⚬⚬⚬

SON OR DAUGHTER OF HE OR SHE

W hen it comes to English grammar, everybody has their own pet peeves.

There are two things "wrong" with that sentence. The most important wrong thing—the absolutely, provably wrong thing—is that it's not true. There are millions who don't give a tinker's damn about English grammar.

The second thing is not necessarily wrong. It's a matter of opinion. The traditional rule would have it "everybody has *his* own pet peeves" or the clumsy "everybody has *his or her* own pet peeves." Amanda J. Smith has suggested this supposedly heretical use of "everybody . . . their" as a reasonable answer to the problem of the now somewhat disreputable use of the masculine pronoun to mean "person."

I think that anyone who is in tune with our times must acknowledge the fact that the *his, his-or-her, their* problem is more pressing than it used to be and that a great many more of us will now accept "everybody . . . their" than would have even 10 years ago.

The mail response to the earlier *Light Refractions* on

this subject was interesting. There are those who think "everybody . . . their" is an abomination and are unwilling even to consider it, and there are some who not only accept it but also defend its use with historical arguments.

J. J. Lamberts, of the English department of Arizona State University, writes, "By all means congratulate Amanda J. Smith of Durham, N.C., for nominating *they, their,* and *them* as pronouns referring to collective and indefinite antecedents.

"But she is scarcely breaking new ground. You will find sporadic instances of this usage in Anglo-Saxon. *The Oxford Dictionary* quotes such worthies as Bishop John Fisher, Henry Fielding, Lord Chesterfield, William Whewell, Walter Bagehot and John Ruskin. . . .

"Actually this usage has always been part of the English language. The great pity is that useful expressions like this one have been schoolmarmed into limbo. It began a couple of hundred years ago when they started teaching attitudes about usage as though they were facts. I suppose no real harm has been done, except that countless people have guilt feelings about their language—which is a stupid state of affairs. And there has been a great increase in inhibited—and therefore bad—writing."

R. L. White, of Sausalito, Calif., tells us that "Jane Austen uses this, I believe regularly, though at the moment I can only turn up one example: 'Everybody has their taste in noises as well as in other matters' [*Persuasion,* 1818]."

We can justify *their* as meaning "his or her," simply because as a matter of historical record, it has stood, in very respectable circumstances, for *his or her. The Oxford English Dictionary* says of *they,* "Often used in reference to a single noun made universal by *every, any, no,* etc., or ap-

plicable to one of either sex (= 'he or she')." That's followed by the instances cited by Mr. Lamberts.

Accepting "everyone . . . they" seems to me a sensible
cracking of an unnecessary dogma. I became firmly convinced of this a few weeks ago. A local NBC show called
"Sunday" was being shot at a large and fairly boisterous
outdoor gathering—an arts-and-crafts show, perhaps—and
I heard approximately this conversation:

> Paul: Where's Jerry?
> Kelly: I don't know. I don't think *he* knows where he is.
> Paul: I don't think *anybody* knows where they are.

Paul is Paul Moyers, a local NBC newsman who's very
likeable but who has only the vaguest notions of what is
grammatical (he recently said, "It's been a difficult time
for he and his family"); and Kelly is Kelly Lang, a young
lady who appears to be very nearly perfect in every way.

The point is, though, that if Paul had, in this instance,
been schoolmarm-grammatical, he'd have said either, "I
don't think anybody knows where he is," which wouldn't
have made his point, or "I don't think anybody knows
where he or she is," which might have left Jerry's gender in
doubt, or he'd have had to search for a totally different
construction. My choice is to accept "I don't think anybody knows where they are" as perfectly all right. It seems
to me that to do otherwise threatens to leave us tongue-
tied.

September 21, 1974

WHO SAYS SO?

In a column called "Linguistic Inanities," I suggested that there was nothing wrong with "aren't I?" as a colloquial substitute for "am I not?"—little knowing that I was attacking quasi-religious dogma by doing so. All hell broke loose, and I felt the slings and arrows of outraged purists. Dozens of them.

One lady asked, "Have you ever in your pseudo-democratic life heard *anyone* say, 'I are not'? I doubt it. Then why on earth should anyone say, 'Aren't I?' Such a senseless, 'inane' manufacture has no justification in logic, grammar, clarity, or colloquiality.

"Whatever you say, 'aren't I?' is certainly *not* almost universal. I've never said it in my puff, never expect to, and never remember anyone saying it. I was surprised to see you dub it 'colloquially respectable.' "

I'm puzzled by that use of "puff," but let it pass. And I won't pick on "never remember anyone saying it," except to suggest that if the lady really doesn't remember hearing "aren't I?," she, as a self-styled "intelligent, college-educated, well-read, lively, middle-aged woman," either has a faulty memory or has led a very sheltered, however lively, life. And it wasn't I, but H. W. Fowler in *Modern English Usage,* who said "aren't I?" was "almost universal" and who dubbed it "colloquially respectable." He and I agree, but the words are his.

A correspondent from Indiana says, " 'Aren't I?' . . . may sound smoother because we are used to it, but wrong it is.

"Neither Webster nor Fowler is an authority on what is 'right' in English. They, the lexicographers, have no judicial powers. Webster and Fowler are reporters. That is right to them which is used by a majority."

I get the impression that many people who can't accept "aren't I?" feel quite at home with sentences like "That is right to them which is used by a majority." *De gustibus.* . . .

But this gentleman has hit upon the crux of the matter: "They, the lexicographers, have no judicial powers."

Who *has* judicial powers?

I think I first wondered about who decided what was "good" English and what was "bad" English when I was taught that we should say "I shall" and "you will" except when there's volition implied, when it becomes "I will" and "you shall."

That struck me as being at once very arbitrary and very fuzzy—a combination that made me uneasy. It wasn't that I was going to worry about what to say; as a practical matter, I was going to say "I'll," except in response to questions such as "Who'll finish up the ice cream?"—in which case I'd say, "*I* will." But the "I shall, you will"—"I will, you shall" problem nagged me, partly because I knew I could never bring myself to say, "I shan't," but, more important, because of the honorable principle of "Who says so?"

Fowler knew that while it has its logic, the English language is not a logical system and is affected by the winds and tides of human behavior.

"Aren't I?" may not be logical, but neither is the verb *to be* logical. In fact, the *Oxford English Dictionary's* opening shot at the verb *be* goes "an irregular and defective verb." The fact is that *be, am, is, are, was,* and *were* came from different parts of the world and found themselves huddled together in the super-accommodating arms of the English language.

"Aren't" is as logical and as illogical for "am not" as "won't" is for "will not." The point is that logic has nothing to do with the case. Sometimes the language is logical, and sometimes it isn't. It's possible for it to be "good-logical" and ugly or "bad-illogical" and beautiful. Every educated person knows that "more perfect" is illogical, "bad" English. But if the Preamble to the Constitution said, "We, the People of the United States, in order to form a more nearly perfect union," I think we'd have lost something.

As with so many questions of language, appropriateness is the key. "Am I not to be accorded the rights and privileges of citizenship as well as its obligations" is much better than "Aren't I to be accorded . . ."; but if you're in a crowd before the deli counter, "I'm after you, aren't I?" is much more fitting than "I'm after you, am I not?"

Asked why, I'm reminded of Louis Armstrong, who, when asked, "What is jazz?" answered, "If you gotta ask what is it, you ain't never gonna know."

If there's one unshakably valid grammatical rule based on logic, it's the one against the double negative, but it's my feeling, probably unverifiable, that the more you know about the English language, the more beautiful you'll find Louis Armstrong's answer, illogical and ungrammatical though it is.

October 5, 1974

CHANGES

L anguage doesn't stand still. To attempt to nail it into place and then say this is how it is and ever will be is a waste of time. It can be done only with a dead language—one that is not used in the daily commerce of living people. Those who would try to pin it down and hold it still remind me of Lieutenant Scheisskopf in *Catch-22*, who wanted to nail his men to a long two-by-four and put swivels in their backs, so that they could win parades.

Most people take the language for granted, of course, but many of us think not only *in* the language but *about* it. We're the ones who are often put off by strange locutions—sometimes with very good reason and sometimes for no reason at all.

A few weeks ago, I was dowsy and getting ready to sleep while the television set was beaming a talk show into the bedroom. Juliet Prowse was being interviewed, and I heard her say, "I don't think too many people realize how much work it takes to be a dancer." As I mentioned, I was drowsy, and my mind was idling out of gear. I wondered how many people would have to realize how much work it takes before Juliet Prowse said, "Hey, that's too many people!" Obviously, what she'd meant was simply "very many people."

I even made a note that I should write something about this misuse of the word *too*.

Well, it was late, and I was tuned in to that pre-sleep frequency in which vague, inchoate propositions can become crystalline absolutes. The next morning, of course, I realized that everyone uses *too* in that sense: "How was the play?" "Not too good." "Is she a good dancer?" "Not too." It's a perfectly acceptable idiom.

I looked into my dictionaries to see what *they* had to say about statements like "I don't think too many people realize." All my dictionaries, even the ancient *Oxford English Dictionary,* agree that *too* can mean "very." Webster's Third says, "Extremely, extravagantly, very." *O.E.D.* gets a bit sexist about it, saying, "Excessively, extremely, exceedingly, very. (Now chiefly an emotional feminine colloquialism.)" It goes on to say, "*None too* . . . 'somewhat insufficiently.' "

I have a feeling that the lexicographers who set down "extremely, extravagantly, very" for *too* sensed at the time that *too* shouldn't really mean that. But it does.

I remember an evening several years ago when my son had gone to the movies. After he had come home, I asked him, "How was the picture?" He shrugged and said, "Not all that good." The expression "not all that good," as far as I know, was fairly new at the time. I'd heard it a few times and been faintly irked by it. I mean I wasn't all that happy with it. So I called Tommy on it. "What do you mean, it wasn't all that good? How good was it supposed to be?" Naturally, he didn't know what I was talking about; so I explained that "all that good" implied an antecedent standard of goodness that the film, in his opinion, had not attained. If, for instance, the film had been plugged as the best of the decade, then "not all that good" would mean, perhaps, that he thought the film was good, but not all *that good*—not, that is, the best of the decade.

I know that there are few greater bores than pedantic conversational nit-pickers, and I've been aware for many years that although Tommy has a certain amount of respect for the proper use of language, he is only a lukewarm grammarian; so I was pleasantly surprised when he saw exactly what I had in mind, grinned, and said, "Hey, that's right! Far out!"

That was a few years ago. Now, the fact is that "all that" is an idiom which has become part of the language, for better or worse. It's not yet listed in many dictionaries, but the editors of *Webster's New World Dictionary* recognize it. Under *that,* they list "all that (Colloq.) 1. so very; used in negative constructions (he isn't *all that* rich)."

And so the changes go on. The only permanent thing is flux. Some of the language becomes archaic as new idioms and new words join the family. Lovers of the language will inevitably deplore certain neologisms, but the neologisms will die only of their own deficiencies and not by fiat. Phrases like "at this point in time," for instance, will fade out because they are ponderous, but "it's not all that good" is not all that bad. It will have a life in the language, and we'll live with it.

October 19, 1974

DAM!

In my September 21 column, I said, "There are millions who don't give a tinker's damn about English grammar," and I reaped a harvest of indignant letters defending the reputation of thinkers. I'll quote the one from Mr. Wilbur W. Young, of Toms River, N.J.:

"For your edification . . . the old-time tinker was sort of a Jack-of-all-Trades (not necessarily given to the use of expletives), among which was that of soldering eave troughs, etc., etc. To confine the molten solder within the vital area, he made a wad of moist clay to form a DAM. When finished with the job the dam was tossed away as a worthless bit of something or other.

"Hence the expression: 'Not worth a tinker's DAM.' And I'm damned if I ever want to see the poor old tinker so demeaned again."

I heard that story long ago, and for years I said "tinker's dam" instead of "tinker's damn." It wasn't until the first time I had to *write* the phrase that I gave it a little more thought. "Not worth a tinker's damn (or dam)" is obviously another way of saying "Not worth a damn." It occurred to me that the whole *dam* business might be a fairy tale.

I turned to Alfred H. Holt's *Phrase and Word Origins* (Dover Publications). Holt says, "In *The Literary Digest* of November 9, 1935, it was categorically stated that this expression refers to the small dam of dough or putty erected by a plumber to keep molten solder from spreading. This, says the *O.E.D.*, is 'an ingenious but baseless conjecture'. . . . It is probably exactly what it purports to be, an idle curse made ineffective by much repetition."

The *Oxford English Dictionary*, in fact, says, "*Not to care*, or *be worth, a tinker's curse or damn,* an intensification of the earlier 'not to care, or be worth, a curse or damn' . . . with reference to the reputed addiction of tinkers to profane swearing." The *O.E.D.* then quotes John MacTaggart's *Scottish Gallovidian Encyclopedia* (1824): "A tinkler's [*sic*] curse she did na care What she did think or say."

In 1874, Edward H. Knight, in *The Practical Dictionary of Mechanics,* wrote, "*Tinker's-dam,* a wall of dough raised around a place which a plumber desires to flood with a coat of solder. The material can be but once used; being consequently thrown away as worthless, it has passed into a proverb, usually involving the wrong spelling of the otherwise innocent word 'dam.' "

I can produce what I think is the ultimate authority on this question. It happens that I have in my library a small and evidently very rare morocco-bound autobiography published privately in 1880. It's the life story of Basil Weaselword, one of the few semi-prominent Victorians ever to have read E. H. Knight's *Practical Dictionary of Mechanics.*

I quote from Weaselword's *Diary of a Semi-Prominent Victorian:*

> It was at a house-party at the estate of Lord Smythe-Fethering-ham that I had one of my outstanding successes, if one will forgive a lapse of modesty. Almost everyone was gathered in the drawing-room, and Lady Bertha had just raised the unpleasant subject of the waistcoat Reggie Taylor had worn at Ascot. I said, perhaps a trifle loudly, "Reggie Taylor's Ascot waistcoat indicates Taylor's tailor's not worth a tinker's damn."
>
> There was a sudden fluttering of fans like the rise of a covey of grouse, and cries of "Have a care, Sir!" "Ladies present!" "Shame!" and that sort of thing.
>
> By great good fortune, I had, only two days previously, read *The Practical Dictionary of Mechanics* by Edward H. Knight. I remembered Knight's having said something about a bit of dough used by tinkers or plumbers, persons of that sort, which he said was called a "tinker's dam." So I explained to Her Ladyship and guests that a "tinker's dam" was in no regard profane, nor had I meant it as such. I told them about *The Practical Dictionary of Mechanics* and related quite convincingly the absurd business about the wad of dough used by plumbers. I thus not only escaped censure, but established myself as a rather more fascinating chap for having displayed familiarity with something so arcane as the mysteries of the plumber's trade.

Though Knight invented the "tinker's dam" theory, it was evidently Basil Weaselword who first put it to practical use.

The fact is that today it's acceptable, if you want to say "tinker's dam," to *say* "tinker's dam." On the other hand, if you mean "tinker's damn," say "tinker's damn."

Say what you mean. That's the important thing.

November 16, 1974

DOUBLE NEGATIVES

T he final paragraph of *Light Refractions* in the October 5 *SR/W* started, "If there's one unshakably valid grammatical rule based on logic, it's the one against the double negative. . . ."

Three erudite gentlemen have taken issue. Michael J. Lahey of West Hartford, Conn., does it in Russian: "Russian is, overall, a more logical language than ours. Yet Russians delight in saying, '*Vwe nee znaeyetye neechovo*,' which translated literally means: 'You do not know nothing.'"

John F. Gummere of Philadelphia says, "If two people are before a magistrate, accused of something, and one of them says, 'I didn't do nothing,' while the other one says, 'I didn't do anything,' everyone in the court knows perfectly well that both are pleading innocent. . . . There's no getting away from this fact. . . . This is not to object to pressing for the sociologically preferable single negative; it's only that we must state, if we record English as it's used by millions of native speakers, that native English employs the double negative all the time and not with the force of a positive."

Professor J. J. Lamberts of Arizona State referred me to the section on "Multiple Negation" in his excellent book *A Short Introduction to English Usage.* He says, "We use the expression 'double negative,' but it would be more accurate to call it a 'multiple negative' or 'redundant negative'. . . . In Old English a sentence with two negatives was permitted. . . . Alfred the Great, the most scholarly Englishman of his generation, wrote this line—here translated literally: 'Therefore shall *no* wise man *not* hate *no* man' [Italics Lamberts's].

". . . It was during the period between Chaucer and Shakespeare that multiple negatives fell out of fashion. Elizabethan writers used them, to be sure, but with less gusto and abandon than writers a few centuries earlier had. Shakespeare's Celia could tell Rosalind, for instance: 'I pray you beare with me, I *cannot* goe *no* further [Italics Lamberts's]."

All right; my unshakably valid grammatical rule has been shaken.

Nevertheless, Mr. Lahey's example from Russian brings to my mind the image, in *The Gulag Archipelago*, of a character who mumbles, "I know nothing," and of a couple of N.K.V.D. huskies brandishing truncheons, insisting, "No, Comrade, you *don't* know nothing!" and utilizing the ways they have of making him talk.

But that's *my* problem. If the Russian way of saying "You know nothing" is "You don't know nothing," my fantasies are irrelevant.

Messrs. Gummere and Lamberts also have a point, but the plea, before a magistrate, of "I didn't do nothing" is readily understood as meaning "I didn't do anything," only if we understand the speaker's style.

Imagine two men, one of whom speaks what has been called "substandard English" and the other of whom speaks polished, professorial prose. If we're familiar with the speech of both men and they both say, "I didn't do nothing," Mr. Substandard and The Professor will be assumed to have meant diametrically opposite things.

Professor Lamberts, whose thoughts on language are generally excellent (meaning, among other things, that he and I usually agree), says, "Prohibition of the multiple negative has always been represented as being completely logical, but . . . the 'rule' that is offered in support states that two negatives make a positive on the grounds, one might presume, that this goes on in mathematics. But . . . the proper formulation is $-a + -a = -2a$. Two negatives, therefore, should be stronger than one."

The comparison of this logic to mathematics, I think, is an error. Try a little language logic on Alfred the Great's sentence, and you'll find that, with his triple negative, he bounces off a positive and back to the desired negative: "Therefore shall no wise man not hate no man."

No wise man shall not hate = All wise men shall hate (positive). Whom shall all wise men hate? No man. Therefore a wise man shall hate no man (negative).

I doubt that Alfred went through that process, but it works out.

Shakespeare's Celia, on the other hand, judged by language logic, says what she evidently doesn't mean. In saying she cannot go no further, she indicates (logically) that she *must* go further. You and I and Rosalind know what she meant, but I think we're making a mistake if we accept the principle that because we understand what she meant, she expressed herself well. In fact, to understand Celia we must assume that she means just the opposite of what her words by themselves express. I don't mind doing that. We do it all the time. But if she'd said, "I cannot go further" or "I can go no further," there would be no question at all of what she meant.

I still think the rule against the double negative is logically sound, and no wise man shall not give me no guff.

November 30, 1974

STOW THAT LINE

I n an earlier *Light Refractions,* I wrote about the use of football metaphors by government officials in describing what our problems are and how we can deal with them. I mentioned that Rep. Gerald Ford, during the hearings for his confirmation as Vice-President, had said, "You don't go out and tackle your quarterback once he has called the play."

I objected then to the football jargon as an inept oversimplification of infinitely complex problems, and I feel even more strongly about it today.

As President, Ford is still bogged down on that soggy gridiron. The world is in very serious trouble in terms of energy and food; the United States is the leading consumer of the earth's raw materials; and the President says, "WIN!" As in the case of so many acronyms, "WIN" was thought of before "Whip Inflation Now," of course. "WIN" is in a class with "Fight, team, fight!" as a political slogan. Back in the time of the N.R.A., "We Do Our Part" at least had the virtue of making a statement.

On October 30, President Ford told the members of his Cabinet to hold the line on spending. "I will hold every department responsible," he said. "You have the ball. You must carry the ball. If you don't score, it's your fault."

Don't get me wrong—as a good American, I like football as much as the next guy. But a serious matter of survival deserves something better than "carry the ball," "score," and "WIN."

This is an excellent example of the use of language in setting the tone of an enterprise. Having established the tone with "WIN," the President blamed inflation largely on the high cost of oil and then went on to tell us about some of the helpful energy-saving hints he'd received from concerned citizens. The fact that there are lots of concerned citizens is good, and the fact that they send their ideas to the President is encouraging. But it's just not true that if we forgo that extra trip to the store, keep our thermostats at 68 degrees, drive at 55 mph, and stop listening to our stereos, the energy crisis will be solved.

The United States, as everyone knows, has been unashamedly—even arrogantly—gluttonous in its use of the earth's raw materials. Richard Nixon actually boasted that this country consumes about 30 percent of the global total.

There's an aura of unreality surrounding our national energy policy, and I'm sure many millions of us are aware of it and are deeply disturbed by it.

Perhaps I'm giving too much importance to the language of the President, but it is a part of the problem, just as it was in Vietnam. We couldn't stop the bombing and withdraw because "you don't punt when you're on your opponent's five-yard line." And now, when many Americans are living on dog food and garbage because of high prices and there's stark famine in parts of Africa and Asia, we are given "WIN" and "you have the ball."

What I'm sure most Americans would like to hear is that we're going to try to put some restrictions on the amount of gasoline our cars should burn; that perhaps we'll stop the insanity of putting up office buildings that are virtually hermetically sealed so that we have to burn enormous amounts of fuel to create an artificial atmosphere in them no matter what the weather; that we might learn to bear a certain amount of heat and cold simply by dressing sensibly; that we'll start thinking of ways to salvage the millions of tons of excellent potential fertilizer that we now cart off to be burned in garbage heaps or flush into sewers from toilets and kitchen garbage-disposal units; that we're going to invest a lot of money and brainpower to utilize energy from the sun, the winds, the tides, and anything else that can't be used up when we've come to depend on it.

Obviously, such radical shifts in direction would be neither easy nor quick, but as the old saying goes (dating way back to July 20, 1969), "If we can put a man on the moon, we should be able to . . . [fill in with the project of your choice]."

At any rate, most of us would welcome someone in government who sounded as though he took our problems —national and global—more seriously than the problems of the Cowboys *vs.* the Redskins.

I think it was Alistair Cooke who referred to the Eisenhower administration as "eight years of government by bromide."

At a time like this, we can't afford government by "We want a touchdown."

January 11, 1975

LENNY

Last night I saw *Lenny,* the extraordinary film about the late Lenny Bruce. Dustin Hoffman's performance is brilliant. So are all the other performances. So is the direction.

So much for my critique.

Enough has been said about the remarkable fact that it was only a few years ago that Lenny Bruce was being arrested and harassed for using certain vulgar words in nightclubs, and that today Dustin Hoffman and several other actors use precisely the same words in a movie that has been given, not an X rating, but an R rating. I think it was probably Lenny Bruce who gave the most telling shove to the old barriers. Once they were breached, they toppled with startling suddenness. There's still debate about the value of the new freedom in film language, and I've met people who refuse to see most of the best new films because they don't want to be subjected to offensive language.

I wasn't a Lenny Bruce fan. I didn't think he was funny. And I didn't think we had to make vulgar language an acceptable part of social discourse, as he apparently did. Not that I object; I just couldn't see it as a worthwhile cause.

Willy-nilly, Lenny won. Today at proper dinner parties you can hear language that was once considered fit principally for military barracks, locker rooms, and pool halls. Whether that's a step in the right direction or not, I don't know, but it's a fact.

The thing Lenny Bruce did that I heartily endorse was to fight for the principle that no one should be jailed for the words he uses, no matter how offensive others might find them. (Obviously, we're not talking here about libel and slander, but of words that are, in themselves, offensive.)

I remember that when Lenny was being arrested for his language, the reports would say he'd been arrested for using an "unprintable" word. It always struck me that no words are actually unprintable and that what was meant was that Bruce had used "a word we'd rather not print, especially since we could get arrested for printing it."

I, for one, am glad that these printable words no longer subject to fine or imprisonment those who, for one reason or another, choose to print them.

There's a scene in *Lenny* that, if accurate, indicates that although Lenny Bruce had great respect for the power of words, there was one very important aspect of communication he didn't understand: that a word is usually dependent on the context for its meaning.

In his nightclub act Lenny asks, "How many niggers are out there?" Then he wanders through the audience saying, "There's a nigger; you look like a kike; I bet you're a

spic; nigger nigger nigger, kike kike kike, spic spic spic."
And he says the reason for doing this is that if we use
the words often enough, they'll lose their power to hurt,
"so that if a little six-year-old kid is called a nigger, it won't
make him cry."

The fact is, of course, that if a little six-year-old kid, or
anyone else, for that matter, is called *anything*—saint, hero,
sweetheart, lover, Rocky, or pegboard—with the necessary
amount of jeering contempt and hatred, it will probably
hurt. In the case of a little six-year-old kid, he'll probably
cry.

The point is that the context determines what is to be
understood.

The first time I heard *honky*, I knew just what it meant.
It was a successor to *ofay* but meaner and more hate-filled.
"Rap" Brown said it on a TV newcast. No one had to
wonder what a honky was, and no one doubted for a
moment that it was an expression of loathing. I think
every white American who watched that newscast was
jolted in one way or another by that word *honky*, spoken
with such vituperation. "Rap" Brown accomplished in
that moment what he'd intended: He shocked, intimidated, angered, saddened—in sum, *affected*—those he was
attacking. If we'd first heard *honky* spoken by, say, Walter
Cronkite—" 'Rap' Brown today revealed a word he uses
for white Americans; he calls them *honkies*"—we'd have
known it to be a racist slur. But the power to jolt is in the
context, not in the word.

These days, I don't find the word *honky* particularly
offensive. It is hatred and blind anger that are frightening
and depressing. Constant repetition of the words *nigger,*

kike, spic, honky, and other disparaging terms will not prevent hate-filled people from making six-year-olds cry, I'm afraid. Only replacing the hate with compassion will do that.

January 25, 1975

AIN'T WE GOT FUN?

Some time ago, I mentioned that there are a good number of people who object to the use of *aren't I?* for *am I not?* This reaction was surprising to me because I would have thought that virtually everyone would be willing by now to accept *aren't I?* as a legitimate phrase. I don't want to belabor the issue, and I bring it up only because the *aren't I?* brouhaha supplied the impetus for this column.

In the *Light Refractions* in which I first broached this relatively unimportant problem (July 13, 1974), I said, "I'd like to see *ain't* given legitimate status." Since then, I've had several letters asking, in essence, "What's wrong with *ain't?*"

In giving second thoughts to *ain't,* I turned to Fowler's *Modern English Usage.* He says, "A(i)n't is merely colloquial, and as used for *isn't* is an uneducated blunder and serves no useful purpose. But it is a pity that *a(i)n't* for *am not,* being a natural contraction and supplying a real want, should shock us as though tarred with the same brush."

The *Oxford English Dictionary*, under *ain't*, refers us to *an't*, where it says, "contraction of *aren't, are not*; colloquially for *am not*; and in illiterate or dialect speech for *is not, has not* (*hadn't*). A later and still more illiterate form is AINT."

So that's what's wrong with *ain't:* It's an illiterate. uneducated blunder. In other words, it's wrong because it's bad English, and the reason it's bad English is that everyone *knows* it's bad English.

I'm going to take back what I said about wanting to see *ain't* given legitimate status. Not that that makes any difference. A word isn't given legitimate status by a vote based on someone's motion, but by a subtle consensus—by its being used by the bulk of literate people for approximately the same purpose. That's what piques linguistic conservatives. Look at that old devil *hopefully*. It seemed as though *hopefully*, almost suddenly, like a mushroom that grows overnight, meant "it is hoped that." Purists can mumble—or even shout—that *hopefully doesn't* mean "it is hoped that." But it's a fact that one of the meanings—by now probably the principal meaning—of *hopefully is* "it is hoped that," and if we don't accept that fact without grumbling, we're wasting time and energy.

Back to *ain't*. I'm glad *ain't* is an illiterate, uneducated blunder. As such it has great value. It is paradoxical. *Ain't* is used from time to time by almost all literate people—in the United States, at least. Here's where I part company with Fowler. When he says *ain't* serves no useful purpose, he is mistaken. The useful purpose of *ain't* is to be "bad" English. It's almost naughty, and it adds levity, color, and fun to the language.

Ian Carmichael, as Lord Peter Wimsey on television, uses *ain't* frequently. I don't remember that Dorothy L. Sayers had Lord Peter saying *ain't* in her novels, although he—rather affectedly, it seems to me—habitually dropped his final *g*s. I've heard that the use of *ain't* was an affectation among upper-class Britons during the period just after the first World War; so I guess Ian Carmichael figured that if Lord Peter was the type to drop his final *g*s, he might as well also say *ain't*. I suppose the use of *ain't* by upper-class Englishmen is comparable to the purchase of patched blue jeans at exorbitant prices in today's high-fashion marts.

(I just realized that there are dozens of Dorothy L. Sayers fans out there who are much better versed in the ways of Lord Peter Wimsey than I, and I'm sure to get baskets of mail pointing out at least half a dozen books in which the original Lord Peter says *ain't*.)

Popular-song lyrics would be immeasurably poorer without *ain't* in all its disrepute: "Ain't Misbehavin'," "I Ain't Got Nobody," "It Ain't Necessarily So," "Ain't It a Shame About Mame." And can you imagine "Don't We Have Fun?" or "Haven't We Got Fun?" or "We don't have a barrel of money"? "It Don't Mean a Thing if You Ain't Got That Swing" has both *it don't* and *you ain't*—both invaluable aids to discovering what is really meant in a sentence employing our old friend the double negative. It's my feeling that "It doesn't mean nothing" means just the opposite of "It don't mean nothing," just as "I haven't done nothing" sounds like a defense against charges of frittering away one's time, while "I ain't done nothing" is certainly a defense against charges of wrongdoing.

Ain't has a very special function in English. It is not easy to define, but I wouldn't want to mess with it. Its glory lies very much in its disrepute.

February 8, 1975

ʊʊʊʊʊʊʊʊʊʊʊʊʊʊʊʊʊʊʊʊʊʊ

WE TAKE YOUR WORD

I can't remember another time when there was as much concern about language as there is today. There have always been books, pamphlets, and lectures on language, but it seems to me the interest has been particularly intense during the past few years, perhaps starting with the public exposure by television of the inarticulate mumblings of some of our college students during the tumultuous Sixties. It was especially hard for us who had been trained by strict grammarians to hear the cream of American youth going, "Uh, yeah, well, ya know, like it's a bummer, right?" Then there was the peculiar jargon of the Watergate principals. We were exposed to the extremes of mindless grunts and convoluted gobbledygook, and I think it made a lot of us more language-conscious than we normally are.

Edwin Newman's *Strictly Speaking* has had a long run on the bestseller lists; there are articles in magazines and newspapers deploring the present state of English; and I've read dozens of letters to the editor that rail against what their writers consider almost criminal misuse of "the language of Shakespeare and Milton."

While I may not agree with all the people who are so upset, I think it's great that they care deeply about the language.

Mr. Warren Weaver, of New Milford, Conn., sent me the text of a speech he had delivered about 15 years ago to the Citizens Advisory Committee of the New York Public Library. It's an excellent speech entitled "Words." Not the least among its many delights for me was that it reminded me of a television show that used to brighten our Sunday afternoons back in the Fifties. I think, though I'm not certain, that the show was called "We Take Your Word." As I remember it, three of the regular panelists were John Mason Brown, Bergen Evans, and Abe Burrows. There were others, and I'm sorry I can't remember who they were. It was one of my favorite shows.

Mr. Weaver reminded me of "We Take Your Word" (I'm assuming that was the name) in this section of his talk:

"How does the meaning of a word change in time? Through usage; and primarily through misusage, since accurate usage would of course always stabilize the status quo. Some, like Dr. Bergen Evans, are very tolerant of this process of change. Language is a living thing, they say; and if enough people make a mistake, it ceases to be a mistake. Others, like John Mason Brown, think that change through misusage tends to degrade the language. As for myself, I am very definitely a John Mason Brown man."

I remember that I, too, was "very definitely a John Mason Brown man." I've shifted my position a bit, I think. I have no right to speak for John Mason Brown, but I can speak for myself as the "John Mason Brown man" of a few years ago.

A few years ago, I would have said, for instance, that *infer* means "to derive from reasoning and evidence" and *imply* means, roughly, "to hint." If someone had insisted that *infer* also means "to hint," I'd have argued that it didn't and that anyone who used *infer* to mean "hint" was simply unaware of the distinction between *imply* and *infer*. That would not alter the fact that *infer* does *not* mean "to hint."

I would no longer argue quite so positively. I'll continue to honor the distinction between the two words and use *imply* for one action and *infer* for its complement. I'll also believe that the person who uses *infer* to mean "hint" is probably unaware of the difference and is almost certainly less interested in words themselves than I wish he were. But I have to admit that as it is very frequently used, *infer* means "hint," and there's no point in my saying it doesn't. *The Random House Dictionary,* while giving "imply" as one of the meanings of *infer,* says "INFER in the sense of *to hint* or *imply* is often criticized, and although it occurs in writing and is frequently heard in speech, schoolteachers and editors nevertheless regard it as a solecism for IMPLY."

As I remember Bergen Evans on "We Take Your Word," he was almost enthusiastic about "change through misusage." At least, as Mr. Weaver says, he was "very tolerant" of it. In fact, though, it does no good to be *in*tolerant of it; but I do feel that there should be some sense of mourning for a "good" word gone "bad." It isn't that good words go bad, of course, but as words change their meanings through misusage, the semantic waters get a bit muddied.

I'm happy about the reawakened interest in language.

This might be a good time for a revival of "We Take Your Word." There are still too many of us who take our words for granted.

February 22, 1975

LANGUAGE PADDING

I'm still getting letters from people who want me to come out strongly against that arch-villain of modern conversation, *ya-know*. Usually, the request goes something like, "Can't we put a stop to the sheer idiocy of people whose whole conversation consists of *ya-know?*"

No, we can't.

Ya-know will fade away—in fact, is now fading away—all by itself. It will never die out entirely, of course, but it is already being replaced by another bit of conversational Styrofoam. These phenomena are an inescapable part of language, I think. To try to put a stop to them is like trying to stop the winds or the waves.

In my early youth, the general function of *ya-know* was served by *well*, or *well, um*, as in "Gerard and I went down to the store and, um, when we went in, well, um, the man asked what we were, um, doing and, well, um," and so on.

I remember reporting as a child on something in school. I said something like, "Well, in this book . . . ," and the teacher interrupted to say, "Tommy, a well is a deep hole in the ground, so when you start a report by saying 'Well,' you're putting yourself in a hole."

I think, though it may be wishful thinking, that even at that age I considered that a pretty inane remark. But it made its point and rendered me tongue-tied for quite a while.

When I regained sufficient self-confidence to speak again, the *wells* and *well, ums* had been intimidated from my speech.

Years later, I picked up *ya-know?* It seems to me that the *ya-know?* that studded the speech of the early Forties was different in character from today's *ya-know.* Back then, it was always a question—not a question that demanded an answer, of course, but its inflection was always interrogatory. Today, more often than not, *ya-know* is a casual statement of unremarkable fact, often with the emphasis on the *ya.*

My father broke me of the *ya-know?* habit by responding, "No, dammit, I *don't know,*" often enough to make me conscious of the fact that I was saying it. This remedy for *ya-know* couldn't be universally successful, because, on the one hand, not many people can say, "No, dammit," with my father's arresting choler, and, on the other hand, not every *ya-know*-sayer is as receptive to that sort of questioning as I was. These days, the average questioner might well get an undeleted expletive for his pains.

During the confused, confusing, frustating late Sixties and early Seventies, we got *like, I mean,* and *ya-know* in various combinations, including *like-I-mean-ya-know.* I suspect that this excess of conversational padding came in response to the needs of young people whose convictions were based on a puzzling complexity of factors pervaded with galling frustrations and could not be glibly articulated.

I'm sure it's been much more than a year since I've heard *like-I-mean-ya-know*. It seems to have yielded to *y'un-nerstan'* and *okay*. I don't know whether or not *okay* has gone national yet, but I predict that if it hasn't, it will, because it now clogs the conversations of young people in both northern and southern California, and for the past few years, unless the news media are mistaken, California has been the trend-setter.

Okay has a strange new role. It goes more or less this way:

"We're talking with Valerie Mallory here. Ms. Mallory is a spokesperson for a group called Extra Sexory Perceptions. Can you tell us something about your group, Valerie?"

"Okay. Extra Sexory Perceptions is actually a factual reality, okay? Okay, it's based on the foundation of heightened consciousness, okay? We feel that once men and women have—okay—transcended their basic biological functions, okay?" And so on.

In recent months I've heard earnest young adults, explaining what have been introduced as important new ideas, pad their discussions with so much okay?-okay kapok that my receiving apparatus has got jammed and I've missed the point entirely.

A few years ago I learned to block out all the *like-I-mean-ya-know* and hear the gist of what was being said. Now I'm back where I started, trying to deal with the new *okay*. By the time I can handle that, something else will come along.

And so it goes, forsooth, i' faith, bejabbers, alors, and there's nothing we can do about it.

Alas.

April 5, 1975

ΙΟΙΟΙΟΙΟΙΟΙΟΙΟΙΟΙΟΙΟΙΟΙΟΙΟΙΟΙΟΙΟΙΟΙΟ

LEARNED LENGTH AND THUNDERING SOUND

D r. Lois DeBakey, professor of scientific communications at Baylor College of Medicine, wrote me a few months ago and enclosed several articles on the sort of work she's doing. She's trying to teach doctors to speak, and particularly to write, English. I wish her great success. She's bucking a very powerful tide, because, as she knows, not just doctors, but possibly the majority of successful Americans—those in positions of responsibility and power —speak and write abominably. I don't mean ungrammatically, but abominably. The businessman who writes, "Kindly be advised that, pursuant to our agreement of the 4th inst., we are shipping . . . ," and the expert in social adjustment who writes, "It is the sense of this department that, in consequence of the proclivity of the subject to withdraw from prolonged social and interpersonal contact, subject should be accorded a maximization of supportive . . . ," are both engaged, not in clear communication, but, at least in part, in mindless bombast.

The trouble is that those who write that way don't seem to be aware that so much of it is drivel. It has become the language of the trade, so to speak. Possibly a social-adjustment expert who wrote good, clear English would be considered a lightweight by his confreres, and the businessman who left out "Kindly be advised that, pursuant to our agreement of the 4th inst.," and said simply, "We are shipping . . . ," would be thought a bit slippery.

Most fields have their own jargon, I suppose—idioms that are understood by few outside the field. Cowboys, pharmacists, astronauts, and bindle stiffs all have their own special terms among the tools of their trade. Generally, these terms are useful. Like a kind of shorthand, they help in communication among members of the same field.

But some fields—especially some of the most respected and well paid—have complex jargons that appear designed less to communicate than to impress.

High-flown, arcane language was once a hallmark of lawyers. An occasional teacher, clergyman, or doctor might develop a degree of periphrastic skill, but lawyers were the ones with the big reputations. Any lawyer who couldn't draw up a contract no layman could understand wasn't much of a lawyer. Now, almost anyone with a half-baked education can and probably will get into the act.

Bureaucrats are probably today's most notorious practitioners of the periphrastic arts. The best work on the subject of bureaucratese has been done by Dr. James H. Boren, founder, president, and chairman of the board of the National Association of Professional Bureaucrats and author of *When in Doubt, Mumble* (Van Nostrand, Reinhold, Co.). Dr. Boren says, "By applying the techniques of orbital dialoguing and giving free rein to policy formulation by subliminal thought processes, the professional bureaucrat can foster the mumbling artistry of the adjustive *status quo*." The *probu* (Boren's word for professional bureaucrat) avoids saying anything he might have to defend later on by learning "to use the qualifying abstractions that spell the difference between routine presentations and neutral masterpieces."

Bureaucratese is to a large extent a manner of speaking designed for expressing as few ideas as possible in as many words as possible in order to give the appearance of dynamic forward thrust while at the same time not rocking the boat.

Dr. DeBakey's doctors haven't the same motivation for speaking badly, by and large. But, as she says, "Physicians . . . usually favor *cholelithiasis* over *gallstones, pyrexia* over *fever, deglutition* over *swallowing*. . . . A patient will probably react differently if told he must be treated for *cephalalgia* than if informed that he will be given something to relieve his *headache*. (He may, however, be unwilling to pay a large fee for medication for the simple headache.)"

Here the implication is that a professional man's services are worth more if they're wrapped in $20 words. I think there's a lot of truth in that. Most people *are* impressed by big words.

The unfortunate consequence is that many a reputation for wisdom has been gained by their use, and many a faulty argument has slipped past our critical guards because we were too buffaloed to ask, "What's that there word mean?"

Oliver Goldsmith, in describing the schoolmaster in *The Deserted Village*, wrote:

While words of learned length, and thundering sound,
Amazed the gazing rustics ranged around;
And still they gazed, and still the wonder grew,
That one small head could carry all he knew.

You can bet that not one of the gazing rustics ever asked for a translation into the good English Dr. DeBakey and the rest of us would appreciate.

April 19, 1975

SPELLING AIRERS

I t was a nice ego trip for me to get a letter from Louis Untermeyer asking my help with a word. He says, "As a confirmed cat-lover, I am an aelurophile. The word is frequently in print—most recently in a *New York Times* account of a cat show—but *aelurophile* isn't in any of the three dictionaries on my shelves. It's not in the *American Heritage* or the *Webster's New World* or even in the omniscient *OED*."

I'd heard the word *aelurophile* and was surprised he couldn't find it; so I checked in all my dictionaries. It's in *Webster's New International,* Third Edition. *WNI III* has been criticized by a great many people, usually because it doesn't stick to the gospel as set forth in *WNI II*. We do tend to resist change. Great as Webster's Second is, the Third is more valuable—to me, at least—in 1975. For instance, it contains *aelurophile*. Where the definition should be, it says "*var. of* AILUROPHILE" and under *ailurophile* it says, "a cat fancier: a lover of cats," which is essentially what it says in *Random House* under *ailurophile*. Evi-

dently, the *Times* likes the *var. sp.*, as we puzzle-makers
often put it.

WNI II doesn't have the word at all. It does tell us
that *Ailuroidea* is "A group of Carnivora including the
cats, civets, and hyenas," and now the thing gets muddled.
Try to follow this: *ailuroidea*, in Webster's Third, is listed
"*syn. of* AELUROIDEA," which is then described as "a super-
family of Carnivora comprising the cats, civets, hyenas. . . ." Back in *WNI II*, *aeluro-* is given as "A combining form from Greek *ailouros*, cat." All of which I find
confusing. Anyway, if it hadn't been for *Webster's Third*,
Mr. Untermeyer and I wouldn't have learned any of this.

Our problem points up one of the major reasons why
there's always a movement afoot somewhere to change
the spelling of the English language into something more
reasonable. You can't look up a word unless you have at
least the first few letters right. If a kid says, "Dad, how
do you spell *fisickle?*" and Dad, hoping to teach self-
reliance as well as spelling, says, "Look it up," he'll just
be jamming a larger wedge into the generation gap.

It might be a good idea to have a dictionary that lists
wrong spellings, as well as variant spellings: NEW-MAT-IC:
wrong spelling of *pneumatic*.

In fact, if you're not busy for the next few months, you
might take that up as a project. *A Dictionary of Incorrect
Spellings*. No library should be without it.

There's one word that's rarely spelled right in any newspaper, and it's a word that's easy to look up, even in a
dictionary of correct spellings. The word is *nauseam*, as in
the expression *ad nauseam*. It's *always* spelled *nauseum*:
"The bleeding-heart liberals still go on insisting *ad nauseum* that the death penalty . . ."

A week ago, I'd have said that *nauseum* is incontrovertibly wrong and that the correct spelling is, was, and always will be *ad nauseam,* because the word is Latin and we can't change a Latin spelling simply because so many people get it wrong. Now, I'm not so sure. If *aeluro-* can be a combining form from Greek *ailouros* (which is worse, since it messes with the beginning of the word, causing problems for *The New York Times,* Louis Untermeyer, and me), *nauseum* could possibly become a *var. sp.* of *nauseam.* Sickening.

I'd be remiss if I didn't mention the one dictionary I own that lists *aelurophile* with that spelling, along with its definition, "a cat-lover." It's *Mrs. Byrne's Dictionary,* by Josefa Heifetz Byrne. The full title, in case you're not familiar with it, is *Mrs. Byrne's Dictionary of Unusual, Obscure, and Preposterous Words.*

I've never met Mrs. Byrne, but the jacket of her dictionary says she's a concert pianist and composer. Her husband, Robert Byrne, edited the dictionary and wrote the introduction. It's a smallish dictionary, claiming "6,000 of the weirdest words in the English language," and I'm grateful to Mrs. Byrne for having compiled it. She lists such words as *quakebuttock* (a coward) and one of my favorites, *philoxenist* (one who is happiest while entertaining strangers), as well as a word consisting of 1,913 letters. Remember poor old *antidisestablishmentarianism?* The 1,913-letter epic is the chemical name for tryptophan synthetase. A protein, thank God; so you'll probably never have to use it or even understand it.

I got *Mrs. Byrne's Dictionary* thanks to Mrs. Brace Paddock, of Pittsfield, Mass., who, at the age of 97, is still solving Double-Crostics. I met Mrs. Paddock's granddaughter, Joan Maxwell, a couple of months ago, and Mrs. Maxwell gave me a copy of the dictionary "for giving my grandmother so much pleasure." I hereby publicly express my gratitude to Mrs. Paddock, Mrs. Maxwell, Mrs. Byrne, and to Mr. Untermeyer, for asking me about *aelurophile* in the first place.

May 17, 1975

SOME INCREDULOUS THINGS

Having read the *Light Refractions* that told of Dr. Lois DeBakey's efforts to teach doctors to speak and write plain, understandable English, my friend Bob Van Scoyk called to ask me if I'd seen a story in the newspaper about a woman in New York—an English actress, he said—who is giving diction lessons to New York cab drivers. I hadn't. I agree with Bob, though, that while it's an excellent idea to make doctors easily understood, teaching New York cab drivers to speak Standard English English—or even Standard American English—is a terrible idea in the first place, and is almost certainly impossible in the second place. New York cab drivers have their own Standard Speech, and we contend if Richard Burton ever obtained a New York hack license, he would have to be taught Standard New York Cabby English. It's one of New York's leading tourist attractions. If I got into a cab at Kennedy Airport and the driver said, "Where would you like to go, sir?" instead of "Weah to, Mac?" I'd look for the Candid Camera, and failing to find it, I'd suspect foul play. As Van Scoyk says, "You shouldn't try to teach

the King's English to guys from Queens." Any New York
cab driver who doesn't lean out his window at least twice
a day and holler, "Wyuncha loin hodda droive?" at other
motorists is not to be trusted.

As I say, I didn't see the story myself, but I'm sure Bob
didn't invent it. I just hope AP or UPI or whoever it was
got it all wrong. Say it ain't da troot.

Speaking of newspapers getting things wrong reminds
me of some clippings sent to me a few weeks ago by
Mrs. Robert E. Ruhoff of Portsmouth, Ohio. Mrs. Ruhoff
has an eagle eye and has found a bouquet of bloopers. She
says, "I am enclosing examples of the use of a word that
is totally incorrect but vaguely resembles the correct word
in sound. . . . I know nothing about the composition of a
newspaper; are all the articles read aloud to the composi-
tors or printers? Could you touch on this in one of your
articles for the *Saturday Review* and, if possible, reassure
us that we are not approaching a time when no one knows
what the other person really means?"

I don't know much about the composition of a newspaper, either, and I've wondered about some of those
typos from time to time.

• From the sports pages of the *Cincinnati Enquirer*,
Mrs. Ruhoff has plucked a bit about Catfish Hunter, who
"even tried to minimize the incredulous deal he made with
the New York Yankees."

• From an ad placed by a jeweler in the same paper, "14
karat gold buckle bracelet . . . many bagels and others
from which to choose."

Come to think of it, I suppose a bagel is a lot of dough
for a ring.

• Also from the *Enquirer*: an eight-column headline

saying, "Theater Magnets Indicted For Fraudulent Use Of Star Names." Samuel Insull was, presumably, an electromagnet.

• The *Columbus Dispatch* had a story about President Ford's speech at the Ohio State University commencement exercises. According to the *Dispatch,* the president of Ohio State, Harold Enarson, said that President Ford "had gone for the juggler." I've never heard Mr. Enarson speak, but I'll bet he didn't say "juggler." There must be presidents of great universities who say, "go for the juggler" and "nucular explosions," but I suspect that the fault lies in the *Dispatch*'s typography. In any case, the President evidently went for the juggler in vein, or we'd have heard more about it.

• My favorite of Mrs. Ruhoff's clippings consists of a pair, one from the *Cincinnati Enquirer* and one from the *Portsmouth Times.* Dick Forbes, writing in the *Enquirer,* quotes Mike Reid, defensive tackle for the Cincinnati Bengals: " 'I would imagine the people in town are sick of hearing of my quest for inner peace,' he said. 'I know exactly what I'm going to do. I've got myself searched out.' " The *Portsmouth Times,* presumably telling the same story, datelined Cincinnati (AP), said, " 'I would imagine the people are sick of hearing of my guest for dinner piece,' said the former Penn State star."

Another "guest for dinner piece" I know of was written by Kaufman and Hart.

Responding to Mrs. Ruhoff's question about "approaching a time when no one knows what the other person really means," I don't think that's a serious danger, except in the area of intentional doublespeak, as practiced by bureaucrats, authorities, and others (like many advertising people) who intentionally frame their language to deceive. The danger lies in language that appears to say something but, on closer examination, says nothing. The misuse of language found by Mrs. Ruhoff in her newspapers really serves more to delight than to baffle.

Like Standard New York City Cabby English, it spices the language.

May 31, 1975